CALCULUS OF
SEVERAL VARIABLES

Harper's Series in Modern Mathematics

I. N. Herstein and Gian-Carlo Rota, Editors

HARPER & ROW, Publishers

New York, Evanston, and London

CALCULUS OF
SEVERAL VARIABLES

by **CASPER GOFFMAN**

Division of Mathematical Sciences
Purdue University

TO EVE

with love and gratitude

CALCULUS OF SEVERAL VARIABLES

Library of Congress Catalog Card Number 65-13472

CONTENTS

PREFACE

There is wide agreement among mathematicians that a modern approach to the calculus of several variables should be a standard part of the training in mathematics and that it should be taught as early as feasible. It is hoped that the present book will be of some help in this connection.

The material of the book has been used at Purdue at two levels of instruction. The first four chapters have formed the contents of a one semester course, meeting three times a week, for juniors. In a similar course for seniors and first year graduate students the entire book has been covered.

Exercises appear at the end of each chapter, numbered according to the section to which the exercise applies. Thus, for example, the exercise at the end of Chapter II, numbered 3.2, is the second exercise for section 3 of that chapter. Many of the exercises are included simply for practice or to fix the ideas discussed in the text; a substantial number test the ingenuity of the student; and a very small number develop the theory beyond, or parallel to, that of the text.

I am grateful to George Pedrick for his careful reading of the book, and for his many suggestions leading to quite a few changes. I would like to thank Richard Darst for calling my attention to a serious error, and James Serrin for his observation that one of the theorems had a peculiar hypothesis, which was changed.

Finally, it is a great pleasure to express my gratitude to Judy Snider for her exceptionally fine typing of the material.

C. G.

CHAPTER I

EUCLIDEAN SPACE

Euclidean space is a vector space in which there is a distance function satisfying certain conditions. In this chapter, we give the definition and main properties of these spaces. We also discuss linear mappings between vector spaces and their properties. We then give a brief treatment of the topology of euclidean spaces.

1. VECTOR SPACE

A **vector space** over the real field R is a set S, together with mappings

$$g: S \times S \to S$$

and

$$v: R \times S \to S.$$

As is customary, the notations $g(x, y) = x + y$ and $v(a, x) = ax$ are used, and the following conditions are assumed to hold:

(**a**) S is an abelian group with respect to the mapping g.
(**b₁**)   For every $x \in S$ and $a, b \in R$, $(ab)x = a(bx)$.
(**b₂**)   For every $x \in S$ and $a, b \in R$, $(a + b)x = ax + bx$.
(**b₃**)   For every $x, y \in S$ and $a \in R$, $a(x + y) = ax + ay$.
(**b₄**)   For every $x \in S$, $1 \cdot x = x$.

We use the symbol θ for the group identity in S. It is easy to show that

$0 \cdot x = \theta$, for every $x \in S$. The group inverse is easily seen to be $(-1)x$ and is written as $-x$. We write $y + (-x)$ as $y - x$.

We give only two examples of vector spaces.

Example A   Let S be the set of n-tuples of real numbers. For

$$x = (x_1, \cdots, x_n) \in S,$$

$$y = (y_1, \cdots, y_n) \in S,$$

let

$$x + y = (x_1 + y_1, \cdots, x_n + y_n).$$

For

$$x = (x_1, \cdots, x_n) \in S$$

and $a \in R$, let

$$ax = (ax_1, \cdots, ax_n).$$

Example B   Let A be any set and X a vector space. Let X^A be the set of all mappings of A into X. For any $f, g \in X^A$, define $f + g$ as the mapping for which

$$(f + g)(\alpha) = f(\alpha) + g(\alpha),$$

for every $\alpha \in A$. For every $f \in X^A$ and $a \in R$, define af as the mapping for which

$$(af)(\alpha) = af(\alpha),$$

for every $\alpha \in A$.

We recall that if A and B are sets, then a mapping $f : A \to B$ is called **injective** if it is one-one, **surjective** if it is onto B, and **bijective** if it is one-one and onto B. Thus, it is injective if $x, y \in A$, $x \neq y$, implies $f(x) \neq f(y)$, and it is surjective if for every $u \in B$ there is an $x \in A$ (perhaps more than one) for which $u = f(x)$.

If S and T are vector spaces, then a mapping

$$f : S \to T$$

is called a **homomorphism** if $f(x + y) = f(x) + f(y)$, for every $x, y \in S$, and if $f(ax) = af(x)$, for every $x \in S$, $a \in R$.

A homomorphism is called an **isomorphism** if it is bijective. Vector spaces S and T are called **isomorphic** if there is an $f : S \to T$ which is an isomorphism.

Remark   A vector space homomorphism is also called a **linear mapping**. Indeed, once we get going, this will be the terminology we shall use.

If S is a vector space, a subset $T \subset S$ is called a **subspace** if $x, y \in T$ implies $x + y \in T$ and if $x \in T$, $a \in R$, implies $ax \in T$.

Let S and T be vector spaces and let $f: S \to T$ be a homomorphism. The set $k(f) \subset S$ consisting of those $x \in S$ for which $f(x) = \theta$ is easily seen to be a subspace of S. It is called the **kernel** of the homomorphism f. The set $r(f) \subset T$ consisting of those $u \in T$, for which there is an $x \in S$ such that $f(x) = u$, is easily seen to be a subspace of T. It is called the **range** of the homomorphism f. Now, f is an isomorphism if and only if $k(f) = \{\theta\}$ and $r(f) = T$.

If S is a vector space, a subset $A \subset S$ is called **linearly independent** if, for every finite set $x^1, \cdots, x^n$ in A, we have

$$a_1 x^1 + \cdots + a_n x^n = \theta$$

only if $a_1 = \cdots = a_n = 0$.

A vector space S is said to have **finite dimension** n if there is a linearly independent set $e_1, \cdots, e_n$ in S such that every $x \in S$ is a linear combination

$$x = a_1 e_1 + \cdots + a_n e_n.$$

The set $e_1, \cdots, e_n$ is then called a **basis** of S.

The definition of dimension is justified by

Theorem 1

If S is finite-dimensional, then every basis of S has the same number of elements.

Proof

Let $e_1, \cdots, e_n$ be a basis of S. Let $y^1, \cdots, y^n, y^{n+1}$ be elements of S. Since

$$y^i = \sum_{i=1}^{n} a_{ij} e_j, \qquad i = 1, \cdots, n + 1,$$

it follows by elementary matrix theory that one of the y^i is a linear combination of the others. Thus every basis of S has no more than n elements. Since no basis has more elements than any other basis, each basis has the same number of elements. ∎

We shall consider only finite-dimensional vector spaces.

Let E be an n-dimensional vector space. Fix a basis $e_1, \cdots, e_n$ in E. Then every $x \in X$ has a representation

$$x = a_1 e_1 + \cdots + a_n e_n.$$

This representation is unique, for if

$$x = b_1 e_1 + \cdots + b_n e_n,$$

then

$$\theta = (a_1 - b_1)e_1 + \cdots + (a_n - b_n)e_n,$$

and the linear independence of $e_1, \cdots, e_n$ implies $a_1 = b_1, \cdots, a_n = b_n$.

In terms of this representation, we have a canonical mapping ϕ of E into the vector space R^n of n-tuples of real numbers (Example A, above) defined by

$$\phi(x) = (a_1, \cdots, a_n).$$

It is an easy matter to show that

$$\phi: E \to R^n$$

is a vector-space isomorphism, and we leave the details to the reader. It follows that any two n-dimensional vector spaces are isomorphic.

2. DEFINITION OF EUCLIDEAN SPACE

Let E be an n-dimensional vector space, $n \geq 1$. We introduce a metric in E which converts it into euclidean n-space. One way in which this may be done is in terms of a scalar product. A **scalar product** is a real number $[x, y]$, associated with each $(x, y) \in E \times E$, such that

(a) $[x, y] = [y, x]$, for every $(x, y) \in E \times E$.
(b) $[x + y, z] = [x, z] + [y, z]$, for every $x, y, z \in E$.
(c) $[ax, y] = a[x, y]$, for every $(x, y) \in E \times E$ and $a \in R$.
(d) $[\theta, \theta] = 0$ and $[x, x] > 0$, for $x \in E$, $x \neq \theta$.

It then follows that $[x, ay] = a[x, y]$ and that $[x, y + z] = [x, y] + [x, z]$.

In the particular case where $E = R^n$, i.e., the vector space of n-tuples of real numbers, if $x = (x_1, \cdots, x_n)$ and $y = (y_1, \cdots, y_n)$, then the function

$$[x, y] = \sum_{i=1}^{n} x_i y_i$$

is easily seen to be a scalar product. The reader will recall that this is the standard scalar product of coordinate (analytic) geometry.

A norm may be introduced in E in terms of the scalar product. For every $x \in E$, the **norm** of x is defined by

$$|x| = [x, x]^{\frac{1}{2}}.$$

Before discussing the properties of the norm, we need

Theorem 2 (Schwarz Inequality)
For every x, y ∈ E, we have

$$[x, y] \le |x| \, |y|.$$

Proof
For every real number a,

$$0 \le [x - ay, x - ay] = [x, x] + a^2[y, y] - 2a[x, y].$$

If $y = 0$, the theorem is obvious. Assume $y \ne 0$, and let $a^2 = [x, x]/[y, y]$ to obtain the desired result. ∎

Remark It is easy to show that equality holds in the Schwarz inequality if and only if there is an a such that either $y = ax$ or $x = ay$.
By Theorem 2 we obtain

Theorem 3
The norm satisfies

(a) $|0| = 0$, $|x| > 0$, if $x \ne 0$,
(b) $|ax| = |a| \, |x|$, for every $x \in E$, $a \in R$,
(c) $|x + y| \le |x| + |y|$, for every $x, y \in E$.

Proof
(a) and (b) are immediate consequences of the definition. For (c),

$$|x + y|^2 = [x + y, x + y] = |x|^2 + 2[x, y] + |y|^2$$
$$\le |x|^2 + 2 |x| \, |y| + |y|^2 = [|x| + |y|]^2,$$

by Theorem 2. ∎

Any function on E, satisfying the properties of Theorem 3, is called a **norm** on E. A metric is defined in terms of a norm by letting the distance $d(x, y)$, for $x, y \in E$, be the nonnegative number

$$d(x, y) = |x - y|.$$

Then

(a) $d(x, x) = 0$, $d(x, y) > 0$, if $x \ne y$,
(b) $d(x, y) = d(y, x)$,
(c) $d(x, y) + d(y, z) \ge d(x, z)$.

The vector space E, together with a scalar product and the concomitant norm and metric, is called euclidean n-space. We show that, for each n, there is essentially one euclidean n-space.

3. ORTHONORMAL BASIS

Two elements x, y, in a euclidean n-space E, are said to be **orthogonal** if $[x, y] = 0$. Clearly, θ is orthogonal to every $x \in E$. An $x \in E$ is called **normal** if $|x| = 1$. A set $A \subset E$ is called **orthonormal** if every $x \in A$ is normal and if, for $x, y \in A$, $x \neq y$, x and y are orthogonal.

We show that

(a)   every orthonormal set is linearly independent,
(b)   there is an orthonormal set consisting of n elements.

For **(a)**, suppose $\{e_1, \cdots, e_m\}$ is an orthonormal set, and let

$$a_1 e_1 + \cdots + a_m e_m = \theta.$$

For each $i = 1, \cdots, m$, we have

$$a_i = [e_i, a_1 e_1 + \cdots + a_m e_m] = [e_i, \theta] = 0,$$

so that

$$a_i = 0.$$

For **(b)**, let $k < n$ and let $e_1, \cdots, e_k$ be orthonormal. Let x^{k+1}, $e_1, \cdots, e_k$ be a linearly independent set. Then let

$$y^{k+1} = x^{k+1} - \sum_{i=1}^{k} [x^{k+1}, e_i] e_i.$$

Now,

$$[y^{k+1}, e_i] = [x^{k+1}, e_i] - [x^{k+1}, e_i] = 0, \qquad i = 1, \cdots, k.$$

Moreover, $y^{k+1} \neq \theta$, in view of its expression. Let

$$e_{k+1} = \frac{y^{k+1}}{|y^{k+1}|}.$$

The set $e_1, \cdots, e_{k+1}$ is orthonormal, and the result follows by induction. We thus have

Theorem 4
Euclidean n-space has an orthonormal basis.

It follows, by Theorem 4, that there is essentially one euclidean n-space, for each n. To show this, we say that euclidean spaces E and F are **isomorphic** if there is a mapping,

$$\phi \colon E \to F,$$

which is a vector-space isomorphism and is such that scalar products are preserved, i.e., $[x, y] = [\phi(x), \phi(y)]$, for every $x, y \in E$. It then follows that distances are preserved, i.e., $d(x, y) = d(\phi(x), \phi(y))$.

Theorem 5

If E and F are euclidean n-spaces, then they are isomorphic.

Proof

Let $e_1, \cdots, e_n$ be an orthonormal basis in E, and $\bar{e}_1, \cdots, \bar{e}_n$ an orthonormal basis in F. Define the linear mapping ϕ by letting

$$\phi(e_i) = \bar{e}_i, \qquad i = 1, \cdots, n.$$

Then for $x = a_1 e_1 + \cdots + a_n e_n$, we have

$$\phi(x) = a_1 \bar{e}_1 + \cdots + a_n \bar{e}_n.$$

Now

$$[a_1 e_1 + \cdots + a_n e_n, b_1 e_1 + \cdots + b_n e_n] = \sum_{i=1}^{n} a_i b_i,$$

and

$$[a_1 \bar{e}_1 + \cdots + a_n \bar{e}_n, b_1 \bar{e}_1 + \cdots + b_n \bar{e}_n] = \sum_{i=1}^{n} a_i b_i,$$

so that scalar products are preserved.

It also follows that, if $x = a_1 e_1 + \cdots + a_n e_n$, then

$$|x| = \left[\sum_{i=1}^{n} a_i^2 \right]^{\frac{1}{2}}$$

and

$$|\phi(x)| = \left[\sum_{i=1}^{n} a_i^2 \right]^{\frac{1}{2}}. \quad \blacksquare$$

Let E again be euclidean n-space. This time, let $e_1, \cdots, e_n$ and $\bar{e}_1, \cdots, \bar{e}_n$ be orthonormal bases in E. Since $e_1, \cdots, e_n$ is a basis,

$$\bar{e}_i = \sum_{j=1}^{n} a_{ij} e_j, \qquad i = 1, \cdots, n.$$

Since

$$[\bar{e}_i, \bar{e}_j] = \delta_{ij},$$

where

$$\delta_{ij} = \begin{cases} 1, & \text{if } i = j, \\ 0, & \text{if } i \neq j, \end{cases}$$

we get

$$\sum_{k=1}^{n} a_{ik} a_{jk} = \delta_{ij}.$$

In other words, the matrix (a_{ij}) is orthogonal.

4. THE DUAL AND SECOND DUAL

In this section, we consider the homomorphisms of an n-dimensional vector space E into the reals (one-dimensional space). Such a mapping of E into R is called a **linear functional**. Thus a linear functional is a mapping,

$$f: E \to R,$$

such that $f(x + y) = f(x) + f(y)$ and $f(ax) = af(x)$.

The set of all linear functionals on E is a vector space, with operations $+ g$ and af defined by

$$(f + g)(x) = f(x) + g(x)$$

and

$$(af)(x) = af(x),$$

for every $x \in E$. This vector space is designated by E^* and is called the **dual** of E.

We first show that E^* is n-dimensional. For this purpose, we note that if $e_1, \cdots, e_n$ is a basis in E, then

(a) every $f \in E^*$ is determined by its values at $e_1, \cdots, e_n$,

(b) for real numbers $a_1, \cdots, a_n$ there is an $f \in E^*$ such that $f(e_i) = a_i$, $i = 1, \cdots, n$.

For **(a)**, let $f \in E^*$, and suppose $f(e_i) = a_i$, $i = 1, \cdots, n$. Now, let $x \in E$. Then x has a unique representation,

$$x = c_1 e_1 + \cdots + c_n e_n.$$

Moreover,

$$f(x) = c_1 f(e_1) + \cdots + c_n f(e_n),$$

which proves the assertion.

For **(b)**, let $a_1, \cdots, a_n$ be real numbers. For

$$x = c_1 e_1 + \cdots + c_n e_n,$$

let

$$f(x) = c_1 a_1 + \cdots + c_n a_n.$$

Then $f(e_i) = a_i$, $i = 1, \cdots, n$, and it is easy to see that f is linear.

In view of **(a)** and **(b)**, there is a bijective mapping of E^* into the vector space R^n of n-tuples of real numbers. Moreover, it is a routine matter to verify that this mapping is a vector-space homomorphism. We thus have

Theorem 6

If E is an n-dimensional vector space, its dual E* is also n-dimensional.

We consider a special basis in E^*. Let $e_1, \cdots, e_n$ be a basis in E. For each $i = 1, \cdots, n$, let $e_i^* \in E^*$ be defined by

$$e_i^*(e_i) = 1, \; e_i^*(e_j) = 0, \qquad i \neq j.$$

It is clear that $e_1^*, \cdots, e_n^*$ is a linearly independent set and so, since E^* is n-dimensional, it is a basis in E^*. It is called the **dual** basis of $e_1, \cdots, e_n$.

Now E^* is a vector space. As such, it has a dual space $E^{**} = (E^*)^*$. This is also n-dimensional. We now show that E^{**} may be identified with E in a natural way.

Let $x \in E$. We associate with x an element $\bar{x} \in E^{**}$ as follows.

For each $x^* \in E^*$, let

$$\bar{x}(x^*) = x^*(x).$$

Then $\bar{x}$ is a mapping on E^* into R. We show $\bar{x} \in E^{**}$. For this,

$$\bar{x}(x^* + y^*) = (x^* + y^*)(x) = x^*(x) + y^*(x) = \bar{x}(x^*) + \bar{y}(x^*),$$

for every $x^*, y^* \in E^*$; and

$$\bar{x}(ax^*) = (ax^*)(x) = ax^*(x) = a\bar{x}(x^*),$$

for every $x^* \in E^*$ and $a \in R$.

We have thus defined a mapping ψ of E into E^{**}, i.e., the mapping which takes each x into the corresponding $\bar{x}$. We show that this mapping is a vector-space homorphism. For every $x, y \in E$ and for every $x^* \in E^*$,

$$\overline{x + y}(x^*) = x^*(x + y) = x^*(x) + x^*(y) = \bar{x}(x^*) + \bar{y}(x^*),$$

so that

$$\overline{x + y} = \bar{x} + \bar{y}.$$

For every $x \in E$ and $a \in R$, and for every $x^* \in E^*$,

$$\overline{ax}(x^*) = x^*(ax) = ax^*(x) = a\bar{x}(x^*),$$

so that

$$\overline{ax} = a\bar{x}.$$

We next show that the mapping ψ is injective. Let $\bar{x} = \psi(x)$ be the group identity in E^{**}. If $x \neq \theta$, there would be an $x^* \in E^*$ so that

$x^*(x) \neq 0$. Then $\bar{x}(x^*) \neq 0$, so that $\bar{x}$ could not be the group identity. Thus $x = \theta$.

Since E and E^{**} are both n-dimensional, we also have that the mapping is bijective. We thus have

Theorem 7

There is a natural isomorphism between E and E^{**}.

5. NORMS IN THE DUAL

Let E again be an n-dimensional vector space. Suppose there is a norm in E. Since R also has a norm, the elements of E^* are linear mappings from one normed vector space into another.

We now show how a norm may be introduced in E^* in terms of the norms in E and in R. Indeed, let

$$|x^*| = \sup \{|x^*(x)|: x \in E, \quad |x| \leq 1\}.$$

That $|x^*| < \infty$ is easy to show if the norm in E is euclidean. It follows for any norm by Theorem 15 given later in this chapter. It is easy to show, and we leave the proof to the reader, that

$$|x^*| = \sup \{|x^*(x)|: x \in E, \quad |x| = 1\}.$$

We verify that we actually have a norm.

(a) θ^* is the zero functional such that $\theta^*(x) = 0$, for every $x \in E$, so that $|\theta^*| = 0$. If $x^* \neq \theta^*$, there is an $x \in E$ such that $|x^*(x)| \neq 0$; hence there is such an x with $|x| = 1$, so that $|x^*| > 0$.

(b) Since $|x^*(ax)| = |a| \, |x^*(x)|$, it follows that $|ax^*| = |a| \, |x^*|$, for every $a \in R$, $x^* \in E^*$.

(c) For every $x^*, y^* \in E^*$,

$$|x^* + y^*| = \sup \{|(x^* + y^*)(x)|: x \in E, \quad |x| = 1\}$$

$$\leq \sup \{|x^*(x)| + |y^*(x)|: x \in E, \quad |x| = 1\}$$

$$\leq \sup \{|x^*(x)|: x \in E, \quad |x| = 1\} + \sup \{|y^*(x)|: x \in E, \quad |x| = 1\}$$

$$= |x^*| + |y^*|.$$

We thus have a norm in E^*. In terms of this norm in E^*, we then obtain a norm in E^{**}. It is interesting that the canonical mapping of E onto E^{**} turns out to be norm preserving. We leave the proof as an exercise for the reader.

Suppose next that E is euclidean n-space, i.e., that the norm is given in terms of a scalar product. We show that E^*, with the norm defined as above, is also euclidean n-space. In order to show this, we must define a scalar product in E^* and show that the corresponding norm is precisely the one defined as above.

Let $e_1, \cdots, e_n$ be an orthonormal basis in E. We define the scalar product first for the dual basis elements. Let $e_1^*, \cdots, e_n^*$ be the dual basis in E^* and let

$$[e_i^*, e_i^*] = 1,$$
$$[e_i^*, e_j^*] = 0, \quad \text{if } i \neq j.$$

By using the Kronecker symbol, this says

$$[e_i^*, e_j^*] = \delta_{ij}.$$

Now the scalar product of $x^* = a_1 e_1^* + \cdots + a_n e_n^*$ and

$$y^* = b_1 e_1^* + \cdots + b_n e_n^*$$

must be defined as

$$[x^*, y^*] = a_1 b_1 + \cdots + a_n b_n.$$

It is a routine matter to verify that this is a scalar product. The corresponding norm is then

$$|x^*| = [x^*, x^*]^{\frac{1}{2}} = \left[\sum_{i=1}^{n} a_i^2 \right]^{\frac{1}{2}},$$

for $x^* = a_1 e_1^* + \cdots + a_n e_n^*$.

We show that the norm defined as above agrees with this norm.

For this, let

$$x = c_1 e_1 + \cdots + c_n e_n \in E.$$

Then

$$x^*(x) = (a_1 e_1^* + \cdots + a_n e_n^*)(c_1 e_1 + \cdots + c_n e_n)$$
$$= a_1 c_1 + \cdots + a_n c_n.$$

By the Schwarz inequality,

$$|x^*(x)| = \left| \sum_{i=1}^{n} a_i c_i \right| \leq \left[\sum_{i=1}^{n} a_i^2 \right]^{\frac{1}{2}} \left[\sum_{i=1}^{n} c_i^2 \right]^{\frac{1}{2}}$$

But $|x| = 1$ implies

$$\sum_{i=1}^{n} c_i^2 = 1.$$

Thus

$$\sup \{|x^*(x)| : x \in E, \quad |x| = 1\} \leq \left[\sum_{i=1}^{n} a_i^2 \right]^{\frac{1}{2}}.$$

But if we take $c_i = \lambda a_i$, $i = 1, \cdots, n$, then the Schwarz inequality becomes an equality so that

$$\sup \{|x^*(x)|: x \in E, \quad |x| = 1\} = \left[\sum_{i=1}^{n} a_i^2\right]^{\frac{1}{2}} = [x^*, x^*]^{\frac{1}{2}}.$$

We have thus shown that the dual of euclidean n-space, with the norm defined at the beginning of this section, is itself euclidean n-space.

6. THE SPACE L(E, F)

In this section, we consider mappings of one vector space into another. Let E be an n-dimensional vector space and F an m-dimensional vector space.

Let $L(E, F)$ be the set of linear mappings of E into F. Thus,

$$f: E \to F$$

is in $L(E, F)$, if $f(x + y) = f(x) + f(y)$, for every $x, y \in E$, and if $f(ax) = af(x)$, for every $x \in E$, $a \in R$. With vector-space operations defined in the standard way for mappings, it is clear that $L(E, F)$ is a vector space. It is a subspace of F^E (Example B, Section 1).

Let $e_1, \cdots, e_n$ be a basis in E and $\bar{e}_1, \cdots, \bar{e}_m$ a basis in F. For each $i = 1, \cdots, n, j = 1, \cdots, m$, we define an element $e_{ij} \in L(E, F)$ by letting

$$e_{ij}(e_i) = \bar{e}_j$$

and

$$e_{ij}(e_k) = \theta, \qquad k \neq i.$$

Then, for every $x = a_1 e_1 + \cdots + a_n e_n \in E$,

$$e_{ij}(x) = a_i \bar{e}_j.$$

We show that the e_{ij}, $i = 1, \cdots, n; j = 1, \cdots, m$ form a basis in $L(E, F)$.

First, suppose

$$\sum_{i=1}^{n} \sum_{j=1}^{m} a_{ij} e_{ij} = \theta.$$

Then

$$\sum_{i=1}^{n} \sum_{j=1}^{m} a_{ij} e_{ij}(e_i) = \theta$$

implies

$$\sum_{j=1}^{m} a_{ij} \bar{e}_j = \theta, \qquad i = 1, \cdots, n,$$

and the linear independence of $\bar{e}_1, \cdots, \bar{e}_n$ implies $a_{ij} = 0$, $i = 1, \cdots, n$; $j = 1, \cdots, m$. The e_{ij} are accordingly linearly independent.

Next, let $f \in L(E, F)$. We show that f is a linear combination of the e_{ij}. We use the fact that f is determined by its values at $e_1, \cdots, e_n$. Suppose

$$f(e_i) = \sum_{j=1}^{m} a_{ij}\bar{e}_j, \qquad i = 1, \cdots, n.$$

Let

$$g = \sum_{i=1}^{n} \sum_{j=1}^{m} a_{ij}e_{ij}.$$

For each $k = 1, \cdots, n$,

$$g(e_k) = \sum_{i=1}^{n} \sum_{j=1}^{m} a_{ij}e_{ij}(e_k) = \sum_{j=1}^{m} a_{kj}\bar{e}_j = f(e_k).$$

Thus, f is g and so it is a linear combination of the e_{ij}. We have completed the proof of

Theorem 8

If E is n-dimensional and F is m-dimensional, then $L(E, F)$ is nm-dimensional and has a basis as described above.

Let $f \in L(E, F)$ and fix a basis $\bar{e}_1, \cdots, \bar{e}_m$ in F. Then, for each $x \in E$,

$$f(x) = f_1(x)\bar{e}_1 + \cdots + f_m(x)\bar{e}_m,$$

where $f_1(x), \cdots, f_m(x)$ are real numbers. Thus the mapping $f: E \to F$ induces, in terms of a basis in F, a set of m mappings $f_i: E \to R$ $i = 1, \cdots, m$.

It is an easy matter to verify that the f_i are linear, so that they are in E^*.

Now, if we fix a basis $e_1, \cdots, e_n$ in E, our mapping acts as

$$f(x_1e_1 + \cdots + x_ne_n) = f_1(x_1e_1 + \cdots + x_ne_n)\bar{e}_1 + \cdots$$
$$+ f_m(x_1e_1 + \cdots + x_ne_n)\bar{e}_m.$$

If we designate a point in F by

$$\bar{x} = \bar{x}_1\bar{e}_1 + \cdots + \bar{x}_m\bar{e}_m,$$

the mapping may be represented as

$$\bar{x}_1 = f_1(x_1e_1 + \cdots + x_ne_n)$$
$$\cdots$$
$$\bar{x}_m = f_m(x_1e_1 + \cdots + x_ne_n),$$

By the linearity of $f_1, \cdots, f_m$, we obtain

$$\bar{x}_1 = x_1 f_1(e_1) + \cdots + x_n f_1(e_n)$$

$$\cdots$$

$$\bar{x}_m = x_1 f_m(e_1) + \cdots + x_n f_m(e_n).$$

If we write $a_{ij} = f_i(e_j)$, $i = 1, \cdots, n$; $j = 1, \cdots, m$, we see that the mapping f is represented by the matrix (a_{ij}).

This means that if $(x_1, \cdots, x_n)$ are the coordinates of x in the basis $e_1, \cdots, e_n$ and if

$$(\bar{x}_1, \cdots, \bar{x}_m) = (f_1(x), \cdots, f_m(x))$$

are the coordinates of $\bar{x}$ in the basis $\bar{e}_1, \cdots, \bar{e}_m$, then we have

$$(a_{ij}) \begin{pmatrix} x_1 \\ \cdot \\ \cdot \\ \cdot \\ x_n \end{pmatrix} = (f_1(x), \cdots, f_m(x)).$$

Suppose there are norms in E and F. We may then define a norm in $L(E, F)$ by letting

$$|f| = \sup \{|f(x)|: x \in E, \quad |x| = 1\}.$$

The proof that this is a norm follows the exact lines of the case $F = R$.

7. OPEN SETS

We now consider the elementary topology of euclidean n-space E. Let $e_1, \cdots, e_n$ be an orthonormal basis in E. Then, for

$$x = x_1 e_1 + \cdots + x_n e_n \in E,$$

we have

$$|x| = \left[\sum_{i=1}^{n} x_i^2 \right]^{\frac{1}{2}}.$$

By an **open sphere** $s(x, r) \subset E$, of center x and radius $r > 0$, we mean the set

$$s(x, r) = \{y: |x - y| < r\}.$$

The corresponding **closed sphere** is the set

$$\bar{s}(x, r) = \{y: |x - y| \leq r\}.$$

A set $S \subset E$ is said to be **open** if, for every $x \in S$, there is an $r > 0$ such that $s(x, r) \subset S$. We list some simple properties of open sets.

(a)   If S_α, $\alpha \in A$, is a set of open sets, then the union

$$S = \bigcup[S_\alpha \colon \alpha \in A]$$

is open.

Proof

Let $x \in S$. Then $x \in S_\alpha$, for an $\alpha \in A$. There is thus an $r > 0$ such that

$$s(x, r) \subset S_\alpha \subset S.$$

(b)   If S_i, $i = 1, \cdots, n$ is a finite set of open sets, then

$$S = \bigcap[S_i \colon i = 1, \cdots, n]$$

is open.

Proof

Let $x \in \bigcap[S_i \colon i = 1, \cdots, n]$. There are numbers $r_1, \cdots, r_n > 0$ with $s(x, r_i) \subset S_i$, $i = 1, \cdots, n$. Let $r = \min(r_1, \cdots, r_n)$. Then $s(x, r) \subset S$.   ∎

However, the intersection of an infinite set of open sets need not be open. For example, let $x \in E$ and let $S_n = s(x, 1/n)$, $n = 1, 2, \cdots$. Then each S_n is open, but the intersection of the sets S_n consists of x alone and is not open.

We next give the structure of an arbitrary open set.

We first define a **closed interval** $I \subset E$ as

$$I = \{x \colon x = x_1 e_1 + \cdots + x_n e_n, \quad a_i \leq x_i \leq b_i, \quad i = 1, \cdots, n\}.$$

A closed interval is called a **closed cubical interval** if there is an $r > 0$ such that $b_i - a_i = r$, $i = 1, \cdots, n$. The number r is called the side of the interval.

An **interior point** of a set $S \subset E$ is an $x \in S$ such that, for some $r > 0$, $s(x, r) \subset S$. We note in passing that a set is open if and only if all of its points are interior points.

Closed intervals I and J are called **nonoverlapping** if $I \cap J$ contains no interior points of I or of J. We shall prove

Theorem 9

If $S \subset E$ is open, then S is the union of a countable set of closed cubical intervals which are pair-wise nonoverlapping.

Proof

Consider a partition of E into pair-wise nonoverlapping closed cubical intervals,

$$I_1^{(1)}, I_2^{(1)}, \cdots, I_n^{(1)}, \cdots,$$

of side 1.

Let

$$J_1^{(1)}, J_2^{(1)}, \cdots, J_n^{(1)}, \cdots,$$

be those of the above intervals which are contained in S and let

$$K_1^{(1)}, K_2^{(1)}, \cdots, K_n^{(1)}, \cdots,$$

be the remaining ones.

Partition each of the intervals

$$K_1^{(1)}, K_2^{(1)}, \cdots, K_n^{(1)}, \cdots,$$

into nonoverlapping closed cubical intervals to obtain

$$I_1^{(2)}, I_2^{(2)}, \cdots, I_n^{(2)}, \cdots,$$

of side $1/2$.

Let

$$J_1^{(2)}, J_2^{(2)}, \cdots, J_n^{(2)}, \cdots,$$

be those of these intervals which are contained in S and let

$$K_1^{(2)}, K_2^{(2)}, \cdots, K_n^{(2)}, \cdots,$$

be the remaining ones.

Continue, in this way, to obtain intervals

$$I_1^{(m)}, \cdots, I_n^{(m)}, \cdots; J_1^{(m)}, \cdots, J_n^{(m)}, \cdots; K_1^{(m)}, \cdots, K_n^{(m)}, \cdots,$$

of side $1/2^{m-1}$, for every m.

The intervals $\{J_j^{(m)}\}$ are nonoverlapping, countable in number, and we have only to show that their union is S.

For this, let $x \in S$. There is a sphere $\sigma(x, r) \subset S$. There is an m such that every cubical interval of side $1/2^{m-1}$ containing x is in $\sigma(x, r)$. Thus, if x is in no $I_n^{(j)}, j < m$, it will be in an $I_n^{(m)}$. ∎

For every set $S \subset E$, we define the **interior** S_i of S to be the union of all open subsets of S. Then S_i is open. It is not hard to see that S_i is the set of interior points of S.

8. CLOSED SETS

We call a set $S \subset E$ **closed** if it is the complement of an open set. It follows immediately that the intersection of a set of closed sets is closed and that the union of a finite set of closed sets is closed.

Closed sets may be studied in terms of the notion of **limit point** of a set. If $S \subset E$, then $x \in E$ is a limit point of S if, for every $r > 0$, $S \cap s(x, r)$ is infinite.

Theorem 10
A set $S \subset E$ is closed if and only if it contains all of its limit points.

Proof
Suppose S contains all of its limit points. Let $x \in CS$. There is an $r > 0$ such that

$$S \cap s(x, r) = \{y_1, \cdots, y_k\},$$

a finite set. Let

$$r' = \min \{r, |x - y_1|, \cdots, |x - y_k|\}.$$

Then $S \cap s(x, r')$ is empty, and CS is open.

Suppose S has a limit point $x \in CS$. Then for every $r > 0$, $S \cap s(x, r)$ is nonempty, so that CS is not open. Hence S is not closed. ∎

Associated with every set $S \subset E$ is a closed set called the **closure** of S; this is the intersection $\bar{S}$ of all closed sets containing S. It is not hard to see that $x \in \bar{S}$ if and only if, for every $r > 0$, $S \cap s(x, r)$ is nonempty. Another simple fact, whose proof we leave to the reader, is that $\bar{S}$ is the union of S and the set of limit points of S.

In particular, closed spheres and closed intervals are examples of closed sets, and we also leave the proof of this to the reader.

9. COMPLETENESS

By a **Cauchy sequence** in E, we mean a sequence $\{x^k\}$ such that, for every $\epsilon > 0$, there is an N such that $m, k > N$ implies $|x^m - x^k| < \epsilon$.

Let

$$x^k = x_1^k e_1 + \cdots + x_n^k e_n, \qquad k = 1, 2, \cdots,$$

where $e_1, \cdots, e_n$ is an orthonormal basis in E. If $\{x^k\}$ is a Cauchy sequence and if, for $\epsilon > 0$, there is N with $|x^m - x^k| < \epsilon$, for every

$m, k > N$, then for each $i = 1, \cdots, n$, we also have $|x^m_i - x^k_i| \leq$ $|x^m - x^k| < \epsilon$. Thus $\{x^k_i\}$ is a Cauchy sequence of reals, for each $i = 1, \cdots, n$.

Conversely, if $\{x^k_i\}$ is a Cauchy sequence of reals for each $i = 1, \cdots,$ n, then for every $\epsilon > 0$, there is an N such that $m, k > N$ implies $|x^k_i - x^m_i| < \epsilon/n$. Thus

$$|x^k - x^m| < \sum_{i=1}^{n} |x^k_i - x^m_i| < \epsilon.$$

This shows that $\{x^k\}$ is Cauchy if and only if $\{x^k_i\}$ is Cauchy for each $i = 1, \cdots, n$.

A sequence $\{x^k\}$ in E is said to **converge** to x if, for every $\epsilon > 0$, there is an N such that $k > N$ implies $|x - x^k| < \epsilon$. In a fashion similar to the above remarks, we may show that $\{x^k\}$ converges to x if and only if each $\{x^k_i\}$, $i = 1, \cdots, n$ converges to some x_i, and then $x = x_1 e_1 + \cdots + x_n e_n$.

We assume that the reader is familiar with the fact that every Cauchy sequence of reals converges. By the above, it then follows that every Cauchy sequence in E converges. We designate this by saying that E is **complete.**

We now draw the following consequence of the completeness of E. Let

$$I_1 \supset I_2 \supset \cdots \supset I_k \supset \cdots$$

be a decreasing sequence of nonempty closed intervals. Then the set

$$\bigcap_{k=1}^{\infty} I_k$$

is nonempty. If the sides of the intervals converge to zero, then the intersection is a single point.

A similar result holds for closed spheres. We leave the proofs to the reader.

10. BOREL COVERING THEOREM

We can prove an important

Theorem 11

If $F_1 \supset F_2 \supset \cdots \supset F_k \supset \cdots$ is a decreasing sequence of nonempty closed sets in E and if F_1 is bounded, then

$$F = \bigcap_{k=1}^{\infty} F_k$$

is nonempty.

Proof

Let I be a closed cubical interval such that $F_1 \subset I$. Partition I into closed cubical intervals, each of side $1/2$ the side of I. At least one of these meets F_k for an infinite set of values of k, and so, since the sequence $\{F_k\}$ is decreasing, it meets all the F_k. By continuing, we obtain a decreasing sequence,

$$I_1 \supset I_2 \supset \cdots \supset I_r \supset \cdots ,$$

of closed cubical intervals with sides converging to zero, such that $I_r \cap F_k$ is nonempty, for every r and k. Let

$$\{x\} = \bigcap_{r=1}^{\infty} I_r,$$

where $\{x\}$ is the set consisting of the one point x. Now, for each k, x is in the closure of F_k, since every open sphere of center x contains an I_r, and so has nonempty intersection with F_k. Since F_k is closed, $x \in F_k$, for every $k = 1, 2, \cdots$. ∎

Our next theorem is

Theorem 12

Every bounded infinite set S in E has at least one limit point.

Proof

Proceed, as in the proof of Theorem 11, to obtain a sequence

$$I_1 \supset I_2 \supset \cdots \supset I_k \supset \cdots$$

of closed cubical intervals, each of side $1/2$ the side of the immediate predecessor, such that $I_k \cap S$ is infinite, for every $k = 1, 2, \cdots$.
Let

$$\{x\} = \bigcap_{k=1}^{\infty} I_k.$$

For every $r > 0$, $\sigma(x, r) \supset I_k$, for some k, so that $\sigma(x, r) \cap S$ is infinite and x is a limit point of S. ∎

This theorem is usually called the Bolzano-Weierstrass theorem.

Corollary

If $\{x^n\}$ is a bounded sequence in E, then it has a convergent subsequence.

We now prove some lemmas which we shall need.

Lemma 1

If $G \subset E$ is open, then for each $x \in G$, there is an open sphere $s(y, r)$, where $y = y_1 e_1 + \cdots + y_n e_n$, the numbers $r, y_1, \cdots, y_n$ are all rational, $x \in s(y, r)$, and $s(y, r) \subset G$.

Proof

There is an $s(x, t) \in G$. Let y be as required and such that $|x - y| < t/3$. Let $t/3 < r < t/2$ be rational. Then $x \in s(y, r)$ and $s(y, r) \subset G$. ∎

We shall call a sphere $s(y, r)$ rational if $y = y_1 e_1 + \cdots + y_n e_n$ and $r, y_1, \cdots, y_n$ are rational.

Lemma 2

If G_α, $\alpha \in A$ is a set of open sets, there is a countable subset G_{α_i}, $i = 1, 2, \cdots$ such that

$$\bigcup G_\alpha = \bigcup G_{\alpha_i}.$$

Proof

The set of open rational spheres is countable. Consider those rational spheres each of which is contained in some G_α. Order them as

$$s_1, s_2, \cdots, s_i, \cdots.$$

For each i, pick α_i so that $G_{\alpha_i} \supset s_i$. Then, by Lemma 1,

$$\bigcup s_i \supset \bigcup G_\alpha.$$

But

$$\bigcup G_{\alpha_i} \supset \bigcup s_i. \quad ∎$$

We now prove the important

Theorem 13 (Borel Covering Theorem)

If $F \subset E$ is a closed bounded set and G_α, $\alpha \in A$, is a set of open sets such that $F \subset \bigcup [G_\alpha : \alpha \in A]$, then $F \subset G_{\alpha_1} \cup \cdots \cup G_{\alpha_m}$ for a certain finite number of the G_α.

Proof

By Lemma 2, we may assume A countable and write the sets G_α, $\alpha \in A$, as $G_1, G_2, \cdots, G_k, \cdots$. Let

$$H_k = G_1 \cup \cdots \cup G_k, \qquad k = 1, 2, \cdots.$$

For each $k = 1, 2, \cdots$, let

$$S_k = F \sim H_k.$$

It is easily seen that each S_k is closed and that

$$S_1 \supset S_2 \supset \cdots \supset S_k \supset \cdots.$$

If each S_k were nonempty, then by Theorem 9,

$$\bigcap_{k=1}^{\infty} S_k$$

would be nonempty. But this contradicts the hypothesis that

$$F \subset \bigcup_{k=1}^{\infty} G_k.$$

Hence there is a k such that S_k is empty, and so $F \subset H_k$. ∎

Closed bounded sets in E are called **compact**.

We give an application of the Bolzano-Weierstrass theorem which we shall need in the next section.

A mapping f on a set $S \subset E$ into the reals is **continuous** at $x \in S$ if, for every $\epsilon > 0$, there is a $\delta > 0$ such that if $y \in S$ and $|x - y| < \delta$, we have $|f(x) - f(y)| < \delta$. Then f is **continuous on S** if it is continuous at every $x \in S$.

Theorem 14

If $S \subset E$ is a closed bounded set and $f: S \to R$ is continuous on S, with $f(x) > 0$ for each $x \in S$, then there is an $\alpha > 0$ such that $f(x) > \alpha$, for each $x \in S$.

Proof

Suppose the conclusion is false. Then, for every k, there is an $x^k \in S$ such that $f(x^k) < 1/k$. The sequence $\{x^k\}$ has a convergent subsequence which converges to $x \in S$, since S is closed. By the continuity of f on S it follows that $f(x) = 0$, which contradicts the hypothesis. ∎

11. EQUIVALENCE OF NORMS

We now use the above fact to show that in an important sense all norms on E are essentially the same as the euclidean norm.

Theorem 15

If $\| \ \|$ is any norm in E and $| \ |$ is the euclidean norm, then there is a $K > 0$ such that $\|x\| \le K|x|$ and $|x| \le K \|x\|$, for every $x \in E$.

Proof

Let $x = x_1 e_1 + \cdots + x_n e_n$. Then

$$\|x\| = \|x_1 e_1 + \cdots + x_n e_n\| \le |x_1| \|e_1\| + \cdots + |x_1| \|e_n\|$$
$$\le n|x| \max \{\|e_1\|, \cdots, \|e_n\|\}.$$

The above implies that $\| \ \|$ is a continuous real function on the euclidean space E. Let

$$S = \{x: |x| = 1\}.$$

Then S is a bounded closed set, and $\|x\| > 0$, for every $x \in S$. There is thus an $\alpha > 0$ so that $\|x\| > \alpha$, for every $x \in S$. It follows that

$$|x| < \frac{1}{\alpha} \|x\|,$$

for every $x \in E$. ∎

For every $r > 0$ the sphere $\sigma(x, r)$, in the norm $\|\ \ \|$, is the set

$$\sigma(x, r) = \{y: \|x - y\| < r\}.$$

A set S is then open if, for every $x \in S$, there is an $r > 0$ such that $\sigma(x, r) \subset S$.

The above theorem yields

Theorem 16

The open sets in E according to the norm $\|\ \ \|$ are the same as those according to the norm $|\ \ |$.

Proof

Let $G \subset E$ be open according to the norm $|\ \ |$ and let $x \in G$. There is an $r > 0$ such that $|y - x| < r$ implies $y \in G$. Suppose $y \in E$ and $\|y - x\| < r/K$, then $|y - x| < r$. Hence G is open according to the norm $\|\ \ \|$. ∎

This important fact allows us to alter the norm to prove mappings are continuous, for example, if such a change is convenient.

12.   CONNECTED OPEN SETS

In this final section, we consider the notion of connected open sets.

We say that a nonempty open set G is **connected** if $G = H \cup J$, where H and J are open and where $H \cap J$ is empty implies H is empty or J is empty.

We say that an open set G is **linearly connected** if $x, y \in G$ implies there is a finite set $x = x_0, x_1, \cdots, x_{k-1}, x_k = y$ in G, such that every line segment $\overline{x_{i-1}x_i} \subset G$, where

$$\overline{uv} = \{w: w = au + (1 - a)v, \quad 0 \leq a \leq 1\}.$$

We prove

Theorem 17

If $G \subset E$ is connected, then it is linearly connected.

Proof

Fix $x \in G$. We say that y is chained to x, if there are $x = x_0, x_1, \cdots,$ $x_{k-1}, x_k = y$ and open spheres $\sigma(x_i, r_i) \subset G$, such that $\sigma(x_{i+1}, r_{i+1}) \cap \sigma(x_i, r_i)$ is nonempty, $i = 0, \cdots, k$. It follows that if y is chained to x, then the line segments $\overline{x_{i-1} x_i}$, $i = 1, \cdots, k$ are in G.

Let S_x be the set of all points chained to x.

Then S_x is open. For if $y = x_k$, then every $z \in \sigma(x_k, r_k)$ is also chained to x, as we see by appending a sphere of center z contained in $\sigma(x_k, r_k)$ and hence in G.

Let $z \in G \sim S_x$. Then z is the center of a sphere in G which does not meet S_x. For, otherwise, z would also be chained to x. Thus, both S_x and $G \sim S_x$ are open sets. They are disjoint, so that one of them is empty. But $x \in S_x$. Thus, for each $x \in G$, $G = S_x$, so that G is linearly connected. ∎

EXERCISES

1.1 Show that $0 \cdot x = \theta$, for every $x \in S$, and that $(-1)x$ is the inverse of x.

1.2 Show that the continuous functions on the real line, with the usual operations of addition and scalar multiplication, form a vector space.

1.3 If X is a vector space, show that there is an infinite set of distinct isomorphisms of X onto itself.

1.4 If S and T are vector spaces and if $f : S \to T$ is a homomorphism, show that the kernel of f is a subspace of S and the range of f is a subspace of T.

1.5 Show that every subspace of S is the kernel of a homomorphism.

1.6 If S is a vector space and $A \subset S$, then the subspace S_A generated by A is the smallest vector space containing A. Show that there is such an S_A.

1.7 Let S be the space of 4-tuples of reals and T the space of triples of reals. Define a homomorphism $f : S \to T$ whose kernel is the vector space generated by $(0, 1, -1, 1)$, $(1, 0, 1, 2)$, and $(2, -1, 3, 1)$.

1.8 Show that A is linearly independent if and only if, whenever B is a proper subset of A, then S_B is a proper subspace of S_A.

1.9 Show that A is a basis of S if and only if $S_A = S$ and $S_B \neq S$, for any proper subset $B \subset A$.

1.10 Furnish those details in the proof of Theorem 1 which were omitted.

1.11 Is $(0, 1, 1, -1)$, $(1, 0, 0, 1)$, $(1, -1, 2, 1)$, $(2, 2, -1, 0)$ a basis for the vector space of 4-tuples of reals?

1.12 Define the isomorphism of the vector space of triples of reals into itself which takes $(1, -1, 0)$ into $(1, 1, 1)$, $(-1, 1, 1)$ into $(1, 1, 0)$; and $(2, 0, 2)$ into $(0, 1, 1)$.

1.13 Give the details of the proof that any two n-dimensional vector spaces are isomorphic.

1.14 Show that the space of Exercise 1.2, is infinite-dimensional.

1.15 Show that the polynomials form a subspace of this space which is not isomorphic to it.

1.16 Show that every infinite-dimensional space has a proper subspace which is isomorphic to it.

2.1 Show that $[x, ay] = a[x, y]$ and $[x, y + z] = [x, y] + [x, z]$, for a scalar product.

2.2 For $E = R^n$, show that

$$[x, y] = \sum_{i=1}^{n} x_i y_i$$

is a scalar product.

2.3 Show that, in the Schwarz inequality, equality holds if and only if either $y = ax$, a real, or $x = 0$.

2.4 Show that the norm in euclidean n-space satisfies the parallelogram law
$$\|x + y\|^2 + \|x - y\|^2 = 2 \|x\|^2 + 2 \|y\|^2.$$

2.5 Show that if a norm satisfies the parallelogram law, it comes from a scalar product.

2.6 For the continuous functions on $[0, 1]$, show that

$$[f, g] = \int_0^1 fg$$

has the properties of a scalar product.

2.7 In the same space, show that the norm
$$|f| = \max \{|f(x)| : x \in [0, 1]\}$$
does not come from a scalar product.

2.8 A convex symmetric set S in a vector space E is one for which $x, y \in S$, $0 \le a, b \le 1$, $a + b = 1$ implies $ax + by \in S$ and for which $x \in S$, $|a| \le 1$ implies $ax \in S$. Show that, for any norm in E, the set for which $|x| < 1$ is a convex, symmetric set.

2.9 Define a metric in E which does not come from a norm.

2.10 Show that the set of sequences, $x = \{x_1, \cdots, x_n, \cdots\}$, for which

$$\sum_{n=1}^{\infty} x_n^2 < \infty,$$

is a vector space with the usual addition and scalar multiplication for sequences.

2.11 Define a scalar product in this space as

$$[x, y] = \sum_{n=1}^{\infty} x_n y_n$$

and prove the Schwarz inequality.

2.12 In the space R^2 of number pairs $x = (x_1, x_2)$, show that the number

$$[x_1^p + x_2^p]^{1/p}$$

has the properties of a norm for any $p \ge 1$.

2.13 If $0 < p < 1$, in the above example, show that the number is not a norm.

2.14 Find a norm in R^2 for which the set, for which $|x| < 1$, is the interior of the diamond in Fig. 1.

2.15 Describe the norm in R^2 for which the interior of the ellipse $x_1^2 + 2x_2^2 = 2$ is the set for which $|x| < 1$.

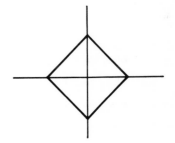

3.1 In R^3, find two orthonormal vectors which generate the same subspace as that generated by $(1, 0, 1)$ and $(-2, 3, 1)$.

Figure 1

3.2 Given the vector $(1, -1, 2)$ in R^3, find two other vectors such that an orthonormal basis is formed by the three vectors.

3.3 In the space of Problem 2.6, find an infinite set of functions which are pair-wise orthogonal.

3.4 In R^2, find an orthogonal matrix which takes the basis $(-1/2, 1/2)$, $(1/2, 1/2)$ onto the basis $(2/5, 1/5)$, $(1/5, -2/5)$.

3.5 In an $n \times n$ matrix, show that the row vectors form an orthonormal set if and only if the column vectors do so.

4.1 Give the details of the proof that E^* is a vector space.

4.2 Complete the proof in the text that E^* is n-dimensional.

4.3 If $E = R^3$, show that every $x^* \in E^*$ is given by a triple (b_1, b_2, b_3) such that the value of x^* at $x = (a_1, a_2, a_3)$ is
$$x^*(x) = a_1 b_1 + a_2 b_2 + a_3 b_3.$$

4.4 Find the dual basis to the basis $(1, 0, 1)$, $(-1, 2, 0)$, $(0, 0, 3)$ in R^3.

4.5 Give the details of the proof that the dual basis in E^* is actually a basis.

5.1 Show that
$$|x^*| = \sup \{|x^*(x)| : x \in E, \quad |x| = 1\}.$$

5.2 Show that the canonical mapping of E onto E^{**} is norm preserving.

5.3 Show that the function
$$[x^*, y^*] = \sum_{i=1}^{n} a_i b_i,$$
introduced in the text, satisfies the conditions of a scalar product.

5.4 Let E be n-dimensional, with basis $e_1, \cdots, e_n$, and for
$$x = a_1 e_1 + \cdots + a_n e_n,$$
let
$$|x| = \max [|a_1|, \cdots, |a_n|].$$
Find the corresponding norm in E^*.

5.5 Then find the corresponding norm in E^{**}.

5.6 If the norm in E is
$$|x| = \left[\sum_{i=1}^{n} |a_i|^p \right]^{1/p}, \quad p > 1,$$
find the corresponding norm in E^*.

5.7 Show that if E is an infinite-dimensional vector space (which may have a countable basis), then E^* does not have a countable basis.

6.1 Show that the e_{ij} defined in the text are elements of $L(E, F)$.

6.2 If $f \in L(E, F)$ and $g \in L(F, K)$, show that $g \circ f \in L(E, K)$.

6.3 If coordinates are prescribed in E, F, and K, show that the matrix which represents $g \circ f$ is the product of the matrices which represent f and g.

6.4 If coordinates are prescribed in E and F, and if, for $x = (x_1, \cdots, x_n) \in E$, $y = (y_1, \cdots, y_n) \in F$, norms are defined by

$$|x| = \max \{|x_1|, \cdots, |x_n|\}$$

and

$$|y| = \max \{|y_1|, \cdots, |y_n|\},$$

what is the corresponding norm in $L(E, F)$?

6.5 If the norms in E and F are euclidean, what is the norm in $L(E, F)$?

6.6 Give details of the proof that $|f| = \sup \{|f(x)| : x \in E, |x| = 1\}$ is a norm in $L(E, F)$.

7.1 Show that the cubical intervals of Theorem 9 can be chosen so that, for every compact K in the open set S, a finite number of the cubical intervals covers K.

7.2 Show that if R is an open rectangle in the plane, R cannot be covered by means of nonoverlapping closed circular disks (i.e., with no interior points in common).

7.3 Show that every open sphere is an open set.

7.4 Show that the interior of a set S is the set of interior points of S.

8.1 Show that $x \in \bar{S}$ if and only if, for every $r > 0$, $s(x, r) \cap S$ is nonempty.

8.2 Show that the closure of S is the complement of the interior of the complement of S.

8.3 Show that $\bar{S}$ is the union of S and the set of limit points of S.

8.4 The boundary of S is defined by

$$\partial S = \bar{S} \cap \overline{CS},$$

where CS is the complement of S. Show that ∂S is closed for every set S.

8.5 Discuss the set $\partial(\partial S)$.

8.6 If S is open, show that $S \cap \partial S$ is empty.

8.7 Show that ∂S is empty if and only if S is either empty or is all of E.

8.8 If S and CS are both closed, show that either $S = E$ or $CS = E$.

8.9 Show that, for any set S, its set of limit points is closed.

9.1 Show that $\{x^k\}$ converges to x if and only if, for every $i = 1, \cdots, n$, $\{x_i^k\}$ converges to x_i.

9.2 Show that the intersection of a decreasing sequence of closed spheres is nonempty, but that the intersection of a decreasing sequence of open spheres can be empty.

10.1 Prove the corollary to Theorem 12.

10.2 Show that if S is compact and $f\colon S \to R$ is continuous, then f is bounded.

11.1 Give a geometric proof of Theorem 15 for the two-dimensional case, using the fact that the set for which $\|x\| \le 1$ is closed, convex, and symmetric, for every norm.

12.1 Define connected sets for sets not necessarily open. Show that if $S \subset E$ is open and connected, dim $E > 1$, and $D \subset E$ is countable, then $S \sim D$ is connected.

CHAPTER II

MAPPINGS AND
THEIR DIFFERENTIALS

This chapter is concerned with nonlinear mappings from one vector space into another. We define continuity and differentiability for such mappings. These definitions are independent of norms as well as of coordinates. The main properties of differentiable mappings are developed.

1. CONTINUOUS MAPPINGS

Let E and F be vector spaces of dimensions n and m, respectively, and let $G \subset E$ be a subset of E. Consider a mapping

$$f: G \to F$$

of G into F.

If we fix bases $e_1, \cdots, e_n$ in E, and $\bar{e}_1, \cdots, \bar{e}_m$ in F, then every $x \in E$ has a unique representation

$$x = x_1 e_1 + \cdots + x_n e_n,$$

and the n-tuple $(x_1, \cdots, x_n)$ of real numbers will be called the **co-ordinates** of x relative to the basis $e_1, \cdots, e_n$. In the same way, there are coordinates in F relative to the basis $\bar{e}_1, \cdots, \bar{e}_m$.

The mapping f may be described in terms of these coordinates. Instead of

$$\bar{x} = f(x),$$

we may write

$$(\bar{x}_1, \cdots, \bar{x}_m) = f(x_1, \cdots, x_n) = (f_1(x_1, \cdots, x_n), \cdots, f_m(x_1, \cdots, x_n))$$

or

$$\bar{x}_1 = f_1(x_1, \cdots, x_n)$$
$$\bar{x}_2 = f_2(x_1, \cdots, x_n)$$
$$\cdots$$
$$\bar{x}_m = f_m(x_1, \cdots, x_n).$$

Thus, the mapping $f: G \to F$ has an associated set of m mappings,

$$f_i: G \to R, \qquad i = 1, \cdots, m,$$

of G into the reals. Of course, if the coordinates in F are changed, then these mappings are changed.

Suppose there are norms in E and F (the same symbol, $|\ |$, will usually be used for every norm involved unless two or more norms are being considered in the same space).

The mapping

$$f: G \to F$$

is said to be **continuous** at $x \in G$ if, for every $\epsilon > 0$, there is a $\delta > 0$ such that $|x - y| < \delta$, $y \in G$ implies $|f(x) - f(y)| < \epsilon$. f is **continuous** on G if it is continuous at every $x \in G$.

Theorem 1

The continuity of f at $x_0 \in G$ is independent of the norms in E and F.

Proof

Suppose f is continuous for norms, written $|\ |$, in E and F, and let there be other norms in E and F, written $\|\ \|$. Then there is a $K > 0$ such that, for every $x \in E$,

$$|x| \leq K \|x\|$$

and, for every $\bar{x} \in F$,

$$\|\bar{x}\| \leq K |\bar{x}|.$$

Let $\epsilon > 0$. There is a $\delta > 0$ such that $|x_0 - y| < \delta$ implies $|f(x_0) - f(y)| < \epsilon/K$. It then follows that $\|x_0 - y\| < \delta/K$ implies $\|f(x_0) - f(y)\| < \epsilon$. ∎

Theorem 2

The mapping $f: G \to F$ is continuous at $x_0 \in G$ if and only if, for every basis $\bar{e}_1, \cdots, \bar{e}_m$ in F, the associated real mappings $f_1, \cdots, f_m$ are continuous.

Proof

Consider the euclidean norm, $|\ |$, in F with orthonormal basis $\bar{e}_1, \cdots, \bar{e}_m$ and coordinates $(\bar{x}_1, \cdots, \bar{x}_m)$. Then, for each $\bar{x} = (\bar{x}_1, \cdots, \bar{x}_m)$, we have

$$|\bar{x}| = [\bar{x}_1^2 + \cdots + \bar{x}_m^2]^{\frac{1}{2}},$$

so that the theorem follows in this case from the inequalities

$$\max [|\bar{x}_1|, \cdots, |\bar{x}_m|] \leq |\bar{x}| \leq |\bar{x}_1| + \cdots + |\bar{x}_m|.$$

The theorem itself follows, by Theorem 1, if we show that continuity of the m coordinate mappings is independent of the coordinate system. We leave the proof of this to the reader. ∎

We shall also be concerned somewhat with directional continuity of f. We assume the domain of f is open.

We say that f is **continuous at** x **in the direction of** $y \in E$ if, for every $\epsilon > 0$, there is a $\delta > 0$, such that $a \in R$, $|a| < \delta$, implies

$$|f(x + ay) - f(x)| < \epsilon.$$

It is possible for f to be continuous at x in every direction without being continuous at x. For example, let E be two-dimensional, with coordinates (x, y) with respect to an orthonormal basis. Let F be one-dimensional. Consider the mapping given by

$$f(x, y) = \begin{cases} 1, & \text{if } x > 0 \text{ and } y = x^2, \\[2mm] 0, & \text{if } x > 0 \text{ and } y = \dfrac{1}{2}x^2, \\[2mm] 0, & \text{if } x > 0 \text{ and } y = 2x^2, \\[2mm] \dfrac{2}{x^2}\left(y - \dfrac{1}{2}x^2\right), & \text{if } x > 0 \text{ and } \dfrac{1}{2}x^2 < y < x^2, \\[2mm] -\dfrac{1}{x^2}(y - 2x^2), & \text{if } x > 0 \text{ and } x^2 < y < 2x^2, \\[2mm] 0, & \text{everywhere else.} \end{cases}$$

Then f is continuous everywhere except at $(0, 0)$. Now, $f(0, 0) = 0$, but for every $\delta > 0$, there is an (x, y) with $f(x, y) = 1$ and $x^2 + y^2 < \delta^2$. Finally, f is continuous at $(0, 0)$ in every direction since, for every a, there is a $\delta > 0$ such that $f(x, ax) = 0$ whenever $x^2 + y^2 < \delta$.

2. DEFINITION OF DIFFERENTIAL

We now discuss differentiable mappings. Again, let E be an n-dimensional normed vector space, let F be an m-dimensional normed vector space, and let $G \subset E$ be open.

A mapping $f: E \to F$ is said to be **differentiable** at $x \in G$ if there is an

$$l_x \in L(E, F),$$

such that, for every $y \in G$,

$$f(y) - f(x) = l_x(y - x) + R(x, y),$$

where

$$\lim_{|y-x| \to 0} \frac{|R(x, y)|}{|y - x|} = 0.$$

The mapping l_x is called the **differential** of f at x. It will also be written as $Df(x)$. It is important that the reader realize that the differential of f at a point x is a linear mapping of E into F. Thus, for every $y, z \in E$, we have

$$[Df(x)](y + z) = [Df(x)](y) + [Df(x)](z),$$

and for every $y \in E$, and real a, we have

$$[Df(x)](ay) = a[Df(x)](y),$$

where, $[Df(x)](y)$, $[Df(x)](z)$, etc., are elements of F.

If f is differentiable at every $x \in G$, we have the differential mapping

$$Df: G \to L(E, F).$$

This mapping takes each $x \in G$ into the differential $Df(x)$ of f at x; we repeat, $Df(x) \in L(E, F)$.

Since the differential of f at x is a linear mapping, it has an evaluation at every $y \in E$. Thus, its evaluation at $y \in E$ is the element $[Df(x)](y) \in F$.

Theorem 3

If f is differentiable at x then for every $y \in E$,

$$[Df(x)](y) = \lim_{a \to 0} \frac{f(x + ay) - f(x)}{a}.$$

Proof

By definition,

$$f(y) - f(x) = [Df(x)](y - x) + R(x, y),$$

where

$$\lim_{|y-x|\to 0} \frac{|R(x, y)|}{|y - x|} = 0.$$

Now,

$$\frac{f(x + ay) - f(x)}{a} = \frac{1}{a}[Df(x)](ay) + \frac{1}{a}R(x, x + ay)$$

$$= [Df(x)](y) + \frac{1}{a}R(x, x + ay).$$

But

$$\lim_{a\to 0} \frac{|R(x, x + ay)|}{|a|} = \lim_{a\to 0} \frac{|y|\,|R(x, x + ay)|}{|ay|} = 0,$$

so that the theorem is proved. ∎

In particular, let $e_1, \cdots, e_n$ be a basis in E, and let $(x_1, \cdots, x_n)$ be the corresponding coordinate system. The evaluations of $Df(x)$ at the basis vectors are called the **partial derivatives** of f with respect to x_i, and are designated by

$$[Df(x)](e_i) = \frac{\partial f}{\partial x_i}(x).$$

We observe that this conforms to standard usage. As a matter of fact, it is possible for

$$\lim_{a\to 0} \frac{f(x + ae_i) - f(x)}{a}, \qquad i = 1, \cdots, n$$

to exist, as will be seen later, without f being differentiable at x. If this limit exists then we say that the partial derivative of f with respect to x_i exists at x, and we again designate it by $\partial f/\partial x_i(x)$. The above remarks then say that, if f is differentiable, its partial derivatives exist and are given by the evaluations of $Df(x)$ at the basis vectors.

Theorem 4

If f is differentiable at x, then the differential is unique.

Proof

Suppose l and m are differentials of f at x. Then

$$f(y) - f(x) = l(y - x) + R(x, y),$$

and

$$f(y) - f(x) = m(y - x) + S(x, y),$$

where

$$\lim_{|y-x|\to 0} \frac{|R(x, y)|}{|y - x|} = \lim_{|y-x|\to 0} \frac{|S(x, y)|}{|y - x|} = 0.$$

Suppose $m \neq l$. Then there is a z such that $(m - l)z \neq 0$. This implies

$$\lim_{a\to 0} \frac{|(m - l)(az)|}{|az|} \neq 0.$$

However, $(m - l)(az) = R(x, x + az) - S(x, x + az)$, so that

$$\lim_{a\to 0} \frac{|(m - l)(az)|}{|az|} \leq \lim_{a\to 0} \frac{|R(x, x + az)| + |S(x, x + az)|}{|az|} = 0.$$

This contradiction establishes the fact that $l = m$. ∎

3. DIFFERENTIABILITY IMPLIES CONTINUITY

We show that if f is differentiable at x then it is continuous there.

Theorem 5

If $f: G \to F$, $G \subset E$ is open, and f is differentiable at $x \in G$, then f is continuous at x.

Proof

We have

$$f(y) - f(x) = l(y - x) + R(x, y)$$

where, for some $\delta > 0$, $|y - x| < \delta$ implies $|R(x, y)| < |y - x|$.

Moreover, there is a K such that

$$|l(x - y)| \leq K|x - y|, \quad \text{for all } y \in E.$$

It follows that

$$|f(y) - f(x)| \leq (K + 1)|y - x|,$$

and the continuity of f at x is implied. ∎

We have shown above that continuity of a mapping from E into F is independent of the norms in E and F. That differentiability is also independent of the norms is just as easy to prove and we leave the proof to the reader. However, in view of its great importance, we now state this fact formally.

Theorem 6
The differentiability of a mapping f on E into F is independent of the norms in E and F.

4. SPECIAL CASES

We now consider some special cases.

Example A In this first example, let $E = F = R$ be one-dimensional spaces. Suppose that $x \in R$ and that

$$f: R \to R$$

is differentiable at x. The differential of f at x is then a linear mapping

$$l_x: R \to R,$$

and for $y \neq x$,

$$f(y) - f(x) = l_x(y - x) + R(x, y),$$

where

$$\lim_{|y - x| \to 0} \frac{|R(x, y)|}{|y - x|} = 0.$$

The vectors in R may be identified with the real numbers. It is then easy to compute that the differential of f at x, evaluated at the vector 1, is simply the customary derivative, $f'(x)$, of f at x. For

$$\frac{f(x + h) - f(x)}{h} = \frac{1}{h} \{l_x(h) + R(x, x + h)\} = l_x(1) + \frac{R(x, x + h)}{h},$$

and

$$\lim_{h \to 0} \frac{f(x + h) - f(x)}{h} = l_x(1).$$

Example B Let E be n-dimensional and let $F = R$. Consider a basis $e_1, \cdots, e_n$ in E, and let $e_i^*, \cdots, e_n^*$ be the dual basis in $E^* = L(E, R)$. Recall that, for each $i = 1, \cdots, n$, e_i^* is that element of E^* for which

$$e_i^*(e_j) = \delta_{ij}.$$

Now, let $f: G \to R$, $G \subset E$ open, be differentiable at $x \in G$ and suppose $l_x \in E^*$ is the differential of f at x. Since $l_x \in E^*$, it may be expressed as

$$l_x = a_1 e_1^* + \cdots + a_n e_n^*.$$

We compute the coefficients $a_1, \cdots, a_n$.

Now

$$f(y) - f(x) = l_x(y - x) + R(x, y) = \sum_{i=1}^{n} a_i e_i^*(y - x) + R(x, y),$$

where

$$\lim_{|y-x| \to 0} \frac{|R(x, y)|}{|y - x|} = 0.$$

Fix $j = 1, \cdots, n$ and let

$$y = x + c e_j.$$

Then

$$f(x + c e_j) - f(x) = \sum_{i=1}^{n} a_i e_i^*(c e_j) + R(x, x + c e_j)$$

$$= c a_j + R(x, x + c e_j).$$

So,

$$\frac{f(x + c e_j) - f(x)}{c} = a_j + \frac{R(x, x + c e_j)}{c}.$$

Since

$$\lim_{c \to 0} \frac{|R(x, x + c e_j)|}{|c e_j|} = \lim_{c \to 0} \frac{|R(x, x + c e_j)|}{|c|} = 0,$$

it follows that

$$a_j = \frac{\partial f}{\partial x_j}(x).$$

Hence, the differential of f at x is given by the formula

$$Df(x) = \sum_{i=1}^{n} \frac{\partial f}{\partial x_i}(x) e_i^*.$$

We also have

$$Df(x) = \sum_{i=1}^{n} [Df(x)](e_i) \cdot e_i^*,$$

since by Theorem 3,

$$[Df(x)](e_i) = \frac{\partial f}{\partial x_i}(x), \qquad i = 1, \cdots, n.$$

Example C Let E be n-dimensional and let F be m-dimensional. Consider a basis $e_1, \cdots, e_n$ in E and a basis $\bar{e}_1, \cdots, \bar{e}_m$ in F. Let $G \subset E$ be open and let

$$f: G \to F$$

be differentiable at $x \in G$. With $(x_1, \cdots, x_n)$ as the coordinates in E with respect to the basis $e_1, \cdots, e_n$ and with $(\bar{x}_1, \cdots, \bar{x}_m)$ as the coordinates in F with respect to the basis $\bar{e}_1, \cdots, \bar{e}_m$, the mapping f may be expressed as

$$\bar{x}_1 = f_1(x_1, \cdots, x_n)$$
$$\cdots$$
$$\bar{x}_m = f_m(x_1, \cdots, x_n).$$

We show that the differentiability of f at x implies the differentiability of the mappings $f_i: G \to R$, $i = 1, \cdots, m$ at x, and conversely. Moreover, the differentials of the mappings f_i are the coordinate mappings of the differential of f, i.e.,

$$Df(x) = (Df_1(x), \cdots, Df_m(x)).$$

Since f is differentiable at x, we have $f(y) - f(x) = l(y - x) + R(x, y)$, where $l \in L(E, F)$ and

$$\lim_{|x-y| \to 0} \frac{|R(x, y)|}{|x - y|} = 0.$$

Let the coordinate mappings corresponding to l be $l_1, \cdots, l_m$ and those corresponding to R be $R_1, \cdots, R_m$. Then $l_i \in E^*$, $i = 1, \cdots, m$, and we have

$$f_i(y) - f_i(x) = l_i(y - x) + R_i(x, y), \qquad i = 1, \cdots, m.$$

Now, for $\bar{x} \in F$, define a norm by

$$\|\bar{x}\| = \sum_{i=1}^{m} |\bar{x}_i|,$$

where $\bar{x} = \bar{x}_1 \bar{e}_1 + \cdots + \bar{x}_m \bar{e}_m$. Then, for each i,

$$|R_i(x, y)| \le \|R(x, y)\| \le K |R(x, y)|,$$

so that f_i is differentiable at x and its differential is l_i, $i = 1, \cdots, m$.

Conversely, suppose each f_i, $i = 1, \cdots, m$ is differentiable at x and the differential of f_i is l_i. Then

$$f_i(y) - f_i(x) = l_i(y - x) + R_i(x, y), \qquad i = 1, \cdots, m,$$

where $l_i \in E^*$ and

$$\lim_{|y-x|\to 0} \frac{|R_i(x, y)|}{|y - x|} = 0.$$

Then,

$$f(y) - f(x) = l(y - x) + R(x, y),$$

where l and R are the mappings of E into F whose corresponding coordinate mappings are l_i and R_i, $i = 1, \cdots, m$, respectively.

But if we use the above norm, $\| \quad \|$, then

$$\|R(x, y)\| = \sum_{i=1}^{m} |R_i(x, y)|,$$

so that

$$\lim_{|y-x|\to 0} \frac{\|R_i(x, y)\|}{|y - x|} = 0.$$

Hence f is differentiable at x, and l is its differential.

It also follows that the matrix of the mapping l is

$$\begin{pmatrix} \dfrac{\partial f_1}{\partial x_1}(x) \cdots \dfrac{\partial f_1}{\partial x_n}(x) \\ \cdots \\ \dfrac{\partial f_m}{\partial x_1}(x) \cdots \dfrac{\partial f_m}{\partial x_n}(x) \end{pmatrix}.$$

This matrix is called the **Jacobian matrix** of the mapping f at x.

5. FUNCTIONS OF CLASS C¹

It is possible for the partial derivatives of f to exist without f being differentiable.

For example, let E be two-dimensional with an orthonormal basis e_1, e_2 and corresponding coordinates x_1, x_2. Let $F = R$ and let f be defined by

$$f(x_1, x_2) = \begin{cases} \dfrac{x_1 x_2}{x_1^2 + x_2^2}, & \text{if } (x_1, x_2) \neq (0, 0), \\ 0, & \text{if } (x_1, x_2) = (0, 0). \end{cases}$$

Then f is differentiable at $(x_1, x_2) \neq (0, 0)$. We leave it to the reader to show this and to compute the differential.

On the other hand, at $(0, 0)$,

$$\frac{\partial f}{\partial x_1} = 0 \quad \text{and} \quad \frac{\partial f}{\partial x_2} = 0,$$

so that if f were differentiable at $(0, 0)$, its differential would have to be the zero functional. We would then have

$$f(x_1, x_2) = R((0, 0), (x_1, x_2)),$$

where

$$\lim_{|(x_1, x_2)| \to 0} \frac{|R((0, 0), (x_1, x_2))|}{|(x_1, x_2)|} = 0.$$

By using the euclidean norm, this implies

$$\lim_{|(x_1, x_2)| \to 0} \frac{x_1 x_2}{(x_1^2 + x_2^2)^{\frac{3}{2}}} = 0.$$

Let $x_1 = x_2$. This implies

$$\lim_{x_1 \to 0} \frac{x_1^2}{2^{\frac{3}{2}} x_1^3} = 0,$$

which is false.

However, continuity of the partial derivatives implies differentiability. Precisely, we have

Theorem 6

If E is an n-dimensional vector space, $G \subset E$ is open, and

$$f: G \to R$$

has continuous partial derivatives in G with respect to the coordinates $(x_1, \cdots, x_n)$ of an orthonormal basis in E, then f is differentiable at every $x \in G$. Moreover, the differential mapping

$$Df: G \to E^*$$

is continuous on G.

Proof

Let $x \in G$ be fixed and let $\epsilon > 0$. There is a sphere $\sigma(x, r)$ such that $y \in \sigma(x, r)$ implies

$$\left| \frac{\partial f}{\partial x_i}(x) - \frac{\partial f}{\partial x_i}(y) \right| < \frac{\epsilon}{n}, \quad i = 1, \cdots, n.$$

Now, let $y \in \sigma(x, r)$. Then the points

$$x^0 = x = (x_1, \cdots, x_n), \quad x^1 = (y_1, x_2, \cdots, x_n),$$

$$x^2 = (y_1, y_2, x_3, \cdots, x_n), \cdots, x^n = y = (y_1, \cdots, y_n)$$

are in $\sigma(x, r)$.

We then have

$$f(y) - f(x) = \sum_{i=1}^{n} \{f(x^i) - f(x^{i-1})\}.$$

By the mean-value theorem for functions of one variable, there is for each $i = 1, \cdots, n$, a

$$y^i = \lambda_i x^i + (1 - \lambda_i)x^{i-1}, \qquad 0 \le \lambda_i \le 1,$$

for which

$$f(x^i) - f(x^{i-1}) = (y_i - x_i)\frac{\partial f}{\partial x_i}(y^i).$$

It follows that

$$f(y) - f(x) = \sum_{i=1}^{n} (y_i - x_i)\frac{\partial f}{\partial x_i}(y^i) = \sum_{i=1}^{n} (y_i - x_i)\frac{\partial f}{\partial x_i}(x) + R(x, y),$$

where

$$|R(x, y)| \le \frac{\epsilon}{n}\sum_{i=1}^{n} |y_i - x_i| < \epsilon |y - x|,$$

where $|y - x|$ is the euclidean norm.

Thus, f is differentiable at x. Since $e_i^*(y - x) = y_i - x_i$, $i = 1, \cdots, n$, the differential of f at x is given by

$$Df(x) = \sum_{i=1}^{n} \frac{\partial f}{\partial x_i}(x)e_i^*.$$

This expression shows that the partial derivatives are the coordinate mappings corresponding to the differential mappings. It follows, since these coordinate mappings are continuous, that the differential mapping is also continuous. ∎

The converse of Theorem 6 also holds. Thus, if the mapping $f: G \to R$ is differentiable and its differential mapping $Df: G \to E^*$ is continuous, then the partial derivatives exist and are continuous. The proof is now easy and is left to the reader.

In the above discussion, any basis (not necessarily an orthonormal basis) may be used. We omit the discussion.

We now define a mapping $f: G \to R$ to be **of class** C^1 if it is differentiable on G and its differential is continuous there. The gist of the above discussion is that f is of class C^1 on G if and only if its partial derivatives exist and are continuous on G.

6. MAPPINGS OF CLASS C^1

We turn to mappings into m-space. Let E be n-dimensional with a basis $e_1, \cdots, e_n$ and corresponding coordinates $x_1, \cdots, x_n$, and let F be m-dimensional with a basis $\bar{e}_1, \cdots, \bar{e}_m$ and corresponding coordinates $\bar{x}_1, \cdots, \bar{x}_m$. Let $G \subset E$ be open and let

$$f: G \to F.$$

In terms of the coordinates, we may write the mapping f as

$$\bar{x}_1 = f_1(x_1, \cdots, x_n)$$
$$\cdots$$
$$\bar{x}_m = f_m(x_1, \cdots, x_n).$$

If f is differentiable at $x \in G$, then its differential at x is given by the matrix

$$\begin{pmatrix} \dfrac{\partial \bar{x}_1}{\partial x_1}(x) \cdots \dfrac{\partial \bar{x}_1}{\partial x_n}(x) \\ \cdots \\ \dfrac{\partial \bar{x}_m}{\partial x_1}(x) \cdots \dfrac{\partial \bar{x}_m}{\partial x_n}(x) \end{pmatrix},$$

which may also be written as

$$\begin{pmatrix} \dfrac{\partial f_1}{\partial x_1}(x) \cdots \dfrac{\partial f_1}{\partial x_n}(x) \\ \cdots \\ \dfrac{\partial f_m}{\partial x_1}(x) \cdots \dfrac{\partial f_m}{\partial x_n}(x) \end{pmatrix},$$

the **Jacobian matrix** of the mapping at x.

Theorem 7

Let E be n-dimensional, F be m-dimensional, $G \subset E$ open, and

$$f: G \to F,$$

such that the partial derivatives of the coordinate functions corresponding to f (in orthonormal coordinates) are continuous. Then f is differentiable on G and its differential is continuous.

Proof

Let $\bar{e}_1, \cdots, \bar{e}_m$ be an orthonormal basis in F. The corresponding coordinate mappings f_i, $i = 1, \cdots, m$ are continuously differentiable so that, for each $x \in G$, there is an $l_{i,x} \in E^*$ such that

$$f_i(y) - f_i(x) = l_{i,x}(y - x) + R_i(x, y), \qquad i = 1, \cdots, m,$$

where

$$\lim_{|y-x| \to 0} \frac{|R_i(x, y)|}{|y - x|} = 0.$$

Moreover, the mapping $x \to l_{i,x}$ is continuous on G to E^*.

But,

$$f(y) = f_1(y)\bar{e}_1 + \cdots + f_m(y)\bar{e}_m,$$
$$f(x) = f_1(x)\bar{e}_1 + \cdots + f_m(x)\bar{e}_m,$$
$$l_x(y - x) = l_{1,x}(y - x)\bar{e}_1 + \cdots + l_{m,x}(y - x)\bar{e}_m,$$

and

$$R(x, y) = R_1(x, y)\bar{e}_1 + \cdots + R_m(x, y)\bar{e}_m,$$

so that we have

$$f(y) - f(x) = l_x(y - x) + R(x, y)$$

and

$$\lim_{|y-x| \to 0} \frac{\|R(x, y)\|}{|y - x|} = 0.$$

Since we have observed that this convergence to zero is independent of norm, the result follows.

That l_x is continuous on G to $L(E, F)$ follows easily from the continuity of the $l_{i,x}$, $i = 1, \cdots, m$, on G to E^*. We leave the proof of this to the reader. ∎

7. COMPOSITION OF DIFFERENTIABLE MAPPINGS

We consider the composition of differentiable mappings in this section. Let E, F, and K be vector spaces whose dimensions are n, m, and k, respectively. Let $G \subset E$ be open and

$$f: G \to F$$

be a mapping of G into F. Now, let $H \subset F$ be open, such that $f(G) \subset H$, and let

$$g: H \to K.$$

Theorem 8

If, with the above arrangement, f is differentiable at $x \in G$ and g is differentiable at $f(x) \in H$, then the mapping

$$g \circ f \colon G \to K$$

is differentiable at x. If l is the differential of f at x and if m is the differential of g at $f(x)$, then $m \circ l$ is the differential of $g \circ f$ at x.

Proof

We must show that

$$(g \circ f)(y) - (g \circ f)(x) = (m \circ l)(y - x) + Q(x, y),$$

where

$$\lim_{|y-x| \to 0} \frac{|Q(x, y)|}{|y - x|} = 0.$$

We first suppose that $f(y) \neq f(x)$. Since g is differentiable at $f(x)$ for $y \in G$, we have

$$(g \circ f)(y) - (g \circ f)(x) = m(f(y) - f(x)) + \bar{R}(f(y), f(x)),$$

where

$$\lim_{|f(y)-f(x)| \to 0} \frac{|\bar{R}(f(y), f(x))|}{|f(y) - f(x)|} = 0.$$

But

$$f(y) - f(x) = l(y - x) + R(x, y),$$

where

$$\lim_{|y-x| \to 0} \frac{|R(x, y)|}{|y - x|} = 0.$$

It follows that

$$(g \circ f)(y) - (g \circ f)(x) = m(l(y) - l(x) + R(x, y)) + \bar{R}(f(x), f(y))$$
$$= m(l(y) - l(x)) + m(R(x, y)) + \bar{R}(f(x), f(y)).$$

There is a K such that, for every $z \in F$,

$$|m(z)| \le K |z|,$$

and there is a K' such that, for every $z \in E$,

$$|l(z)| \le K' |z|.$$

$$\frac{|\bar{R}(f(y), f(x))|}{|y - x|} = \frac{|\bar{R}(f(y), f(x))|}{|f(y) - f(x)|} \frac{|f(y) - f(x)|}{|y - x|}$$
$$= \frac{|\bar{R}(f(y), f(x))|}{|f(y) - f(x)|} \cdot \frac{|l(y) - l(x) + R(x, y)|}{|y - x|}$$
$$\le \frac{|\bar{R}(f(y), f(x))|}{|f(y) - f(x)|} \cdot \left[K' + \frac{|R(x, y)|}{|y - x|} \right].$$

We thus have

$$\frac{|m(R(x, y)) + \bar{R}(f(x), f(y))|}{|y - x|} \leq \frac{K |R(x, y)|}{|y - x|}$$

$$+ \frac{|\bar{R}(f(y), f(x))|}{|f(y) - f(x)|} \cdot \left[K' + \frac{|R(x, y)|}{|y - x|} \right].$$

Let $\epsilon > 0$. There is a $\delta' > 0$ such that $|y - x| < \delta'$ implies

$$\frac{|R(x, y)|}{|y - x|} < \min \left[\frac{\epsilon}{2K}, K' \right].$$

There is an $\eta > 0$ such that $|z - f(x)| < \eta$ implies

$$\frac{|\bar{R}(z, f(x))|}{|z - f(x)|} < \frac{\epsilon}{4K'}.$$

Moreover, there is a $\delta'' > 0$ such that $|y - x| < \delta''$ implies

$$|f(y) - f(x)| < \eta.$$

Let $\delta = \min (\delta', \delta'')$. Now $|y - x| < \delta$ implies

$$\frac{|m(R(x, y)) + \bar{R}(f(y), f(x))|}{|y - x|} < \epsilon.$$

But this implies that $g \circ f$ is differentiable at x and its differential is $m \circ l$.

Suppose now that $f(y) = f(x)$. Then

$$l(y - x) = -R(x, y),$$

so that

$$(m \circ l)(y - x) = -m(R(x, y))$$

and

$$(g \circ f)(y) - (g \circ f)(x) = (m \circ l)(y - x) + m(R(x, y)),$$

since both sides of the equation are zero. The result follows since

$$\lim_{|y - x| \to 0} \frac{|m(R(x, y))|}{|y - x|} = 0. \quad \blacksquare$$

If we consider coordinate systems in E, F, and K we may express this result in terms of the Jacobian matrices of the mappings f and g.

Let $e_1, \cdots, e_n$ be a basis in E, let $\bar{e}_1, \cdots, \bar{e}_m$ be a basis in F, and let $\bar{\bar{e}}_1, \cdots, \bar{\bar{e}}_k$ be a basis in K. The corresponding coordinates are then designated by $x_1, \cdots, x_n, \bar{x}_1, \cdots, \bar{x}_m$, and $\bar{\bar{x}}_1, \cdots, \bar{\bar{x}}_k$.

The differential of f at x has then the Jacobian matrix

$$
A = \begin{pmatrix} \dfrac{\partial \bar{x}_1}{\partial x_1} & \cdots & \dfrac{\partial \bar{x}_1}{\partial x_n} \\ \cdots & & \\ \dfrac{\partial \bar{x}_m}{\partial x_1} & \cdots & \dfrac{\partial \bar{x}_m}{\partial x_n} \end{pmatrix},
$$

the differential of g at $f(x)$ has the Jacobian matrix

$$
B = \begin{pmatrix} \dfrac{\partial \bar{\bar{x}}_1}{\partial \bar{x}_1} & \cdots & \dfrac{\partial \bar{\bar{x}}_1}{\partial \bar{x}_m} \\ \cdots & & \\ \dfrac{\partial \bar{\bar{x}}_k}{\partial \bar{x}_1} & \cdots & \dfrac{\partial \bar{\bar{x}}_k}{\partial \bar{x}_m} \end{pmatrix}
$$

and the differential of $g \circ f$ at x has the Jacobian matrix

$$
C = \begin{pmatrix} \dfrac{\partial \bar{\bar{x}}_1}{\partial x_1} & \cdots & \dfrac{\partial \bar{\bar{x}}_1}{\partial x_n} \\ \cdots & & \\ \dfrac{\partial \bar{\bar{x}}_k}{\partial x_1} & \cdots & \dfrac{\partial \bar{\bar{x}}_k}{\partial x_n} \end{pmatrix}.
$$

But Theorem 8 tells us that

$$
C = BA = \begin{pmatrix} \sum_{i=1}^{m} \dfrac{\partial \bar{\bar{x}}_1}{\partial \bar{x}_i} \dfrac{\partial \bar{x}_i}{\partial x_1} & \cdots & \sum_{i=1}^{m} \dfrac{\partial \bar{\bar{x}}_1}{\partial \bar{x}_i} \dfrac{\partial \bar{x}_i}{\partial x_n} \\ \cdots & & \\ \sum_{i=1}^{m} \dfrac{\partial \bar{\bar{x}}_k}{\partial \bar{x}_i} \dfrac{\partial \bar{x}_i}{\partial x_1} & \cdots & \sum_{i=1}^{m} \dfrac{\partial \bar{\bar{x}}_k}{\partial \bar{x}_i} \dfrac{\partial \bar{x}_i}{\partial x_n} \end{pmatrix}.
$$

We thus have

$$
\frac{\partial \bar{\bar{x}}_r}{\partial x_s} = \sum_{i=1}^{m} \frac{\partial \bar{\bar{x}}_r}{\partial \bar{x}_i} \frac{\partial \bar{x}_i}{\partial x_s}, \qquad \begin{array}{l} r = 1, \cdots, k, \\ s = 1, \cdots, n. \end{array}
$$

We now give two examples, one direct and simple, the other more complicated.

Example A Let

$$
f: R^3 \to R^2
$$
$$
g: R^2 \to R^3.
$$

In terms of coordinate systems, the mappings f and g have representations

$$u = u(x, y, z),$$

$$f:$$

$$v = v(x, y, z),$$

and

$$r = r(u, v),$$

$$g: s = s(u, v),$$

$$t = t(u, v),$$

and we obtain

$$\frac{\partial r}{\partial x} = \frac{\partial r}{\partial u}\frac{\partial u}{\partial x} + \frac{\partial r}{\partial v}\frac{\partial v}{\partial x},$$

$$\frac{\partial r}{\partial y} = \frac{\partial r}{\partial u}\frac{\partial u}{\partial y} + \frac{\partial r}{\partial v}\frac{\partial v}{\partial y},$$

$$\frac{\partial r}{\partial z} = \frac{\partial r}{\partial u}\frac{\partial u}{\partial z} + \frac{\partial r}{\partial v}\frac{\partial v}{\partial z},$$

and similar expressions for the partial derivatives of s and t with respect to x, y, and z.

Example B In this example, we suppose the coordinates are already given and that we have

$$u = u(x, y),$$

$$v = v(x, y, u),$$

$$w = w(x, u, v).$$

The problem is to find the mappings which are involved.

(a) There is first a mapping from the space in which the coordinates are (x, y) into the space whose coordinates are (x, y, u).
(b) Then there is a mapping from the space in which the coordinates are (x, y, u) into the space whose coordinates are (x, u, v).
(c) Finally, there is a mapping from this space in which the coordinates are (x, u, v) into the space whose coordinate is w.

We thus have the sequence of mappings

$$(x, y) \rightarrow (x, y, u) \rightarrow (x, u, v) \rightarrow w.$$

These three mappings have Jacobian matrices

$$
\begin{pmatrix} 1 & 0 \\ 0 & 1 \\ u_x & u_y \end{pmatrix}, \qquad
\begin{pmatrix} 1 & 0 & 0 \\ 0 & 0 & 1 \\ v_x & v_y & v_u \end{pmatrix}, \qquad \text{and} \qquad
(w_x w_u w_v),
$$

respectively. The composite mapping which takes the space with coordinates (x, y) into the one with coordinate w is then given by the matrix

$$
(w_x w_u w_v) \begin{pmatrix} 1 & 0 & 0 \\ 0 & 0 & 1 \\ v_x & v_y & v_u \end{pmatrix} \begin{pmatrix} 1 & 0 \\ 0 & 1 \\ u_x & u_y \end{pmatrix}
$$

$$
= (w_x + w_v v_x \quad w_v v_y \quad w_u + w_v v_u) \begin{pmatrix} 1 & 0 \\ 0 & 1 \\ u_x & u_y \end{pmatrix}
$$

$$
= (w_x + w_v v_x + w_u u_x + w_v v_u u_x \quad w_v v_y + w_u u_y + w_v v_u u_y).
$$

This yields the formulas

$$
\frac{\partial w}{\partial x} = \frac{\partial w}{\partial x} + \frac{\partial w}{\partial v}\frac{\partial v}{\partial x} + \frac{\partial w}{\partial u}\frac{\partial u}{\partial x} + \frac{\partial w}{\partial v}\frac{\partial v}{\partial u}\frac{\partial u}{\partial x}
$$

and

$$
\frac{\partial w}{\partial y} = \frac{\partial w}{\partial v}\frac{\partial v}{\partial y} + \frac{\partial w}{\partial u}\frac{\partial u}{\partial y} + \frac{\partial w}{\partial v}\frac{\partial v}{\partial u}\frac{\partial u}{\partial y}.
$$

It may be necessary to remark that in the first equation $\partial w/\partial x$ on the left side refers to the composite mapping $(x, y) \to w$, and $\partial w/\partial x$ on the right side refers to the mapping $(x, u, v) \to w$, so that they are not the same.

8. HIGHER DIFFERENTIALS

We give a brief discussion of higher differentials.

Let E be n-dimensional, F be m-dimensional, $G \subset E$ be open, and

$$
f: G \to F
$$

be differentiable on G. Then, for every $x \in G$, the differential $Df(x)$ of f at x is an element of $L(E, F)$.

We recall that if $e_1, \cdots, e_n$ is a basis in E and $\bar{e}_1, \cdots, \bar{e}_m$ is a basis in F, and if $x_1, \cdots, x_n$ and $\bar{x}_1, \cdots, \bar{x}_m$ are the corresponding coordinates in E and F, respectively, then a linear mapping $l \in L(E, F)$ is given by

$$\bar{x}_1 = a_{11}x_1 + \cdots + a_{1n}x_n$$
$$\cdots$$
$$\bar{x}_m = a_{m1}x_1 + \cdots + a_{mn}x_n.$$

In the particular case of the differential, $Df(x)$, of f at $x \in G$, the corresponding matrix is the Jacobian matrix

$$\begin{pmatrix} \dfrac{\partial f_1}{\partial x_1}(x) & \cdots & \dfrac{\partial f_n}{\partial x_n}(x) \\ \cdots & & \\ \dfrac{\partial f_m}{\partial x_1}(x) & \cdots & \dfrac{\partial f_m}{\partial x_n}(x) \end{pmatrix},$$

where

$$\bar{x}_1 = f_1(x_1, \cdots, x_n)$$
$$\cdots$$
$$\bar{x}_m = f_m(x_1, \cdots, x_n)$$

are the coordinate mappings which correspond to the mapping f.

Now, consider the mapping

$$Df: G \rightarrow L(E, F)$$

which takes each $x \in G$ into the differential $Df(x)$ of f at x.

If we consider the basis e_{ij}, $i = 1, \cdots, n$; $j = 1, \cdots, m$, in $L(E, F)$, where e_{ij} is the mapping which satisfies

$$e_{ij}(e_i) = \bar{e}_j$$
$$e_{ij}(e_k) = 0, \quad k \neq i,$$

the coordinate mappings corresponding to Df, in the coordinates corresponding to this basis, are the nm partial derivatives

$$\frac{\partial f_j}{\partial x_i}(x), \quad i = 1, \cdots, m; j = 1, \cdots, m.$$

Suppose the mapping Df is differentiable on G. Its differential $D^2f = D(Df)$ is called the second differential of f. It is a mapping

$$D^2f: G \to L(E, L(E, F)).$$

Its Jacobian matrix is

$$\begin{pmatrix} \dfrac{\partial^2 f_1}{\partial x_1 \, \partial x_1}(x) & \cdots & \dfrac{\partial^2 f_i}{\partial x_1 \, \partial x_j}(x) & \cdots & \dfrac{\partial^2 f_m}{\partial x_1 \, \partial x_n}(x) \\ \cdots & & & & \\ \dfrac{\partial^2 f_1}{\partial x_n \, \partial x_1}(x) & \cdots & \dfrac{\partial^2 f_i}{\partial x_n \, \partial x_j}(x) & \cdots & \dfrac{\partial^2 f_m}{\partial x_n \, \partial x_n}(x) \end{pmatrix},$$

where

$$\frac{\partial^2 f_i}{\partial x_r \, \partial x_s}(x) = \frac{\partial}{\partial x_r}\left(\frac{\partial f_i}{\partial x_s}(x)\right)$$

are called the second partial derivatives of f_i.

If the second partial derivatives of the f_i, $i = 1, \cdots, m$, exist and are continuous, then D^2f is continuous.

We may define differentials of all orders by induction. We only note that the third differential D^3f is a mapping

$$D^3f: G \to L(E, L(E, L(E, F))).$$

EXERCISES

1.1 If $f: G \to F$ has continuous coordinate mappings for given coordinate systems in E and F, show that the same holds for any other coordinate systems in E and F.

1.2 If $f: G \to F$, $G \subset E$, is continuous, and $g: H \to K$, $f(G) \subset H \subset F$, is continuous, show that $g \circ f: G \to K$ is continuous.

1.3 Show that the continuous mappings on an open $G \subset E$ into F form a vector space.

1.4 Discuss the continuity of the mapping $f: R^2 \to R^2$ given by

$$u = \frac{x}{x^2 + y^2},$$

$$v = \frac{y}{x^2 + y^2},$$

when $(x, y) \neq (0, 0)$; and $u = 0$, $v = 0$, when $(x, y) = (0, 0)$.

1.5   Discuss the continuity of the mapping given by

$$u = \frac{x}{(x^2 + y^2)^\alpha},$$

$$v = \frac{y}{(x^2 + y^2)^\alpha},$$

when $(x, y) \neq (0, 0)$; and $u = 0$, $v = 0$, when $(x, y) = (0, 0)$ for values of $\alpha > 0$.

1.6   Prove that there is a continuous mapping which takes each open square $(-a, a) \times (-a, a)$ onto the open disk $|x| < a$.

1.7   Show that a continuous mapping takes connected sets into connected sets.

1.8   Show that a continuous mapping takes compact sets into compact sets.

1.9   Show that the function f given by

$$f(x, y) = \sin xy$$

is continuous everywhere.

1.10 Prove that every polynomial in 2 variables is continuous.

1.11 A mapping f on a closed interval is quasi-linear if f is continuous and the interval may be divided into finitely many subintervals on each of which f is linear. Show that every continuous mapping is the uniform limit of a sequence of quasi-linear mappings.

1.12 Obtain an analogous result for functions of 2 variables.

1.13 Show that if

$$[a, b] = \bigcup_{n=1}^{\infty} S_n,$$

there is an n and a subinterval $[c, d] \subset [a, b]$ such that S_n is dense in $[c, d]$.

1.14 Let $f: R^2 \to R$ be such that, for each x, f is continuous in y, and for each y, f is continuous in x. Show that if S is zero on a dense set, it is identically zero.

2.1   Find the differential of the mapping given by

$$u = x^2 - 2y$$
$$v = x^2 - 2xy$$
$$w = 3x^2y - 2y$$

at the point $(3, -2)$.

2.3 Show that

$$D(f + g) = Df + Dg$$

at every point where both differentials exist.

2.4 Give an example of a differentiable mapping on R^3 into R^3 whose differential has two-dimensional range at a point.

2.5 Give an example of a differentiable mapping on R^3 into R^3 whose differential is zero at one point and has two-dimensional range everywhere else.

2.6 Give an example of a differentiable mapping on R^3 into R^3 whose differential has one-dimensional range at a point on the unit sphere and has three-dimensional range everywhere else.

3.1 Give an example of a continuous mapping of R^2 into R^2 whose differential does not exist at an infinite set of points.

3.2 Show that every continuous

$$f: R \to R$$

is the uniform limit of C^1 functions.

3.3 Show that if $f: R^2 \to R$ is continuous, then for every n, there are C^1 functions $f_1, \cdots, f_k; g_1, \cdots, g_k$, such that

$$\left| f(x, y) - \sum_{i=1}^{k} f_i(x) g_i(y) \right| < \frac{1}{n},$$

whenever $x^2 + y^2 \leq n^2$.

4.1 Give an example of a differentiable function for which

$$\lim_{|x-y| \to 0} \frac{|R(x, y)|}{|x - y|^n} = 0,$$

but

$$\lim_{|x-y| \to 0} \frac{|R(x, y)|}{|x - y|^{n+1}} \neq 0,$$

for a given n and x.

4.2 Give an example of a differentiable $f: R^2 \to R$, such that $f(x, y) \neq 0$ if $(x, y) \neq (0, 0), f(0, 0) = 0$ and, for every n,

$$\lim_{|x| \to 0} \frac{|R(0, x)|}{|x|^n} = 0.$$

4.3 If f and g are differentiable functions on R^n into R, show that fg is differentiable and find an expression for its differential.

4.4 Write the equation of the tangent plane to a surface

$$z = f(x_1, \cdots, x_n)$$

at a point on the surface, given that f is differentiable.

4.5 Define the normal to a surface. Write the equation of the normal.

5.1 Prove that the function $\sin xy$ is continuously differentiable. Compute its differential at $(1, \pi/2)$.

5.2 Discuss the existence of the differential and partial derivatives of

$$f(x, y) = \begin{cases} \dfrac{x^3 - xy^2}{x^2 + y^2}, & \text{when } (x, y) \neq (0, 0), \\ 0, & \text{when } (x, y) = (0, 0). \end{cases}$$

6.1 Show that if $f: G \to R$ is continuously differentiable, then the partial derivatives of f exist and are continuous on G.

6.2 Carry out the argument of Theorem 6 for a basis which is not necessarily orthonormal.

7.1 If $w = f(x, u, v)$, $u = g(x, y)$, and $v = h(y, z)$, find

$$\frac{\partial w}{\partial x}, \quad \frac{\partial w}{\partial y}, \quad \frac{\partial w}{\partial z}.$$

7.2 If $u = f(x - ct) + g(x + ct)$, show that

$$c^2 \frac{\partial^2 u}{\partial x^2} = \frac{\partial^2 u}{\partial t^2}.$$

7.3 If $z = y f(x^2 - y^2)$, show that

$$y \frac{\partial z}{\partial x} + x \frac{\partial z}{\partial y} = \frac{xz}{y}.$$

7.4 f is said to be positive homogeneous of degree n, if $f(tx, ty) = t^n f(x, y)$. If f is positive homogeneous of degree n, evaluate

$$x \frac{\partial f}{\partial x} + y \frac{\partial f}{\partial y}.$$

8.1 Give an example of a discontinuous function all of whose partial derivatives, pure and mixed, of all orders, exist.

8.2 Express the Laplacian

$$\frac{\partial^2 u}{\partial x^2} + \frac{\partial^2 u}{\partial y^2} + \frac{\partial^2 u}{\partial z^2}$$

in cylindrical coordinates.

8.3 Express the Laplacian in spherical coordinates.

8.4 If $x = f(u, v)$, $y = g(u, v)$, satisfies the Cauchy-Riemann equations

$$\frac{\partial f}{\partial u} = \frac{\partial g}{\partial v}, \qquad \frac{\partial f}{\partial v} = -\frac{\partial g}{\partial u},$$

and if $w = w(x, y)$, show that

$$\frac{\partial^2 w}{\partial u^2} + \frac{\partial^2 w}{\partial v^2} = \left(\frac{\partial^2 w}{\partial x^2} + \frac{\partial^2 w}{\partial y^2}\right)\left[\left(\frac{\partial f}{\partial u}\right)^2 + \left(\frac{\partial f}{\partial v}\right)^2\right].$$

8.5 Write the equation of the directional derivative of $f(x, y)$, along a curve $y = g(x)$, in terms of the partial derivatives of f.

8.6 Do the same for functions of n variables.

CHAPTER III

MAPPINGS INTO
THE REALS

We prove Taylor's theorem for mappings on n-space into the reals. This theorem says that if a function f has continuous differentials of order up to and including k, and if its differential of order $k + 1$ exists, then f may be "approximated" locally by a polynomial of degree k. We then apply this theorem to the theory of relative maxima and minima. We then define an integral and develop its properties.

1. TAYLOR'S THEOREM FOR 1 VARIABLE

We first consider a mapping
$$f: R \to R$$
on the reals into the reals.

We recall that the derivative of f at x is the value of the differential $Df(x)$ at the unit vector. The kth derivative is the value of $D^k f(x)$ at the unit vector. We shall use the notation
$$f'(x), f^2(x), \cdots, f^k(x), \cdots$$
for the successive derivatives of f at x, if they exist. We now prove

Theorem 1

If $f: G \to R$, with $G \subset R$ open, is $k + 1$ times differentiable on G, then if $x \in G$, $y \in G$, there is a ξ between x and y such that

$$f(y) = f(x) + (y - x)f'(x) + \cdots + \frac{(y - x)^k}{k!} f^k(x) + \frac{(y - x)^{k+1}}{(k + 1)!} f^{k+1}(\xi).$$

Proof

We consider the function F on the closed interval $[x, y]$ into the reals (if $y < x$, the interval is $[y, x]$), given by

$$F(t) = f(y) - f(t) - (y - t)f'(t) - \cdots - \frac{(y - t)^k}{k!} f^k(t)$$

$$- \frac{(y - t)^{k+1}}{(k + 1)!} \left[f(y) - f(x) - f'(x)(y - x) - \cdots \right.$$

$$\left. - \frac{f^k(x)}{k!} (y - x)^k \right] \frac{(k + 1)!}{(y - x)^{k+1}}.$$

Then

$$F(x) = F(y) = 0.$$

By Rolle's theorem, there is a $\xi \in (x, y)$ such that $F'(\xi) = 0$. But this simply says that

$$0 = - \frac{f^{k+1}(\xi) \cdot (y - \xi)^k}{k!}$$

$$+ \frac{(y - \xi)^k}{k!} \left[f(y) - f(x) - \cdots - \frac{f^k(x)}{k!} (y - x)^k \right] \cdot \frac{(k + 1)!}{(y - x)^{k+1}}$$

or

$$f(y) = f(x) + (y - x)f'(x) + \cdots + \frac{(y - x)^k}{k!} f^k(x)$$

$$+ \frac{(y - x)^{k+1}}{(k + 1)!} f^{k+1}(\xi). \quad \blacksquare$$

For the case $k = 0$, this is the mean-value theorem

$$f(y) = f(x) + (y - x)f'(\xi), \qquad x < \xi < y.$$

For the case $k = 1$, it is

$$f(y) = f(x) + (y - x)f'(x) + \frac{(y - x)^2}{2!} f^2(\xi), \qquad x < \xi < y.$$

2. TAYLOR'S THEOREM FOR n VARIABLES

We obtain an analogous result for a mapping $f: E \to R$, where E is n-dimensional. As a matter of fact, we obtain this result by considering the function f restricted to the line segment joining the points x and y under consideration, and thus reducing the problem to the one-dimensional case.

Let E be n-dimensional, $G \subset E$ be open, and $f: G \to R$ be $k + 1$ times differentiable on G. Let $x \in G$, $y \in G$ be such that the line segment joining x and y is in G. On that portion of the line passing through x and y which is in G, we have the function g of a real variable t defined by

$$g(t) = f(x_1 + t(y_1 - x_1), \cdots, x_n + t(y_n - x_n)),$$

where we have chosen a coordinate system in E and where $(x_1, \cdots, x_n)$ are the coordinates of x and $(y_1, \cdots, y_n)$ the coordinates of y.

With partial derivatives evaluated at the point $x + t(y - x)$, we obtain

$$g'(t) = \sum_{i=1}^{n} \frac{\partial f}{\partial x_i} (y_i - x_i)$$

$$g^2(t) = \sum_{i,j=1}^{n} \frac{\partial^2 f}{\partial x_i \, \partial x_j} (y_i - x_i)(y_j - x_j)$$

$$\cdots$$

$$g^k(t) = \sum_{i_1, \cdots, i_k = 1}^{n} \frac{\partial^k f}{\partial x_{i_1} \cdots \partial x_{i_k}} (y_{i_1} - x_{i_1}) \cdots (y_{i_k} - x_{i_k})$$

$$g^{k+1}(t) = \sum_{i_1, \cdots, i_{k+1} = 1}^{n} \frac{\partial^{k+1} f}{\partial x_{i_1} \cdots \partial x_{i_{k+1}}} (y_{i_1} - x_{i_1}) \cdots (y_{i_{k+1}} - x_{i_{k+1}}).$$

By Theorem 1,

$$g(1) = g(0) + g'(0) + \frac{1}{2!} g^2(0) + \cdots + \frac{1}{k!} g^k(0) + \frac{1}{(k+1)!} g^{k+1}(\tau),$$

where $0 < \tau < 1$.

By substitution, this yields

Theorem 2

If E is n-dimensional, $G \subset E$ is open, and $f: G \to R$ is $k + 1$ times differentiable on G; then if the closed segment joining x to y is in G, there is a ξ on this line segment such that

$$f(y) = f(x) + \sum_{i=1}^{n} \frac{\partial f(x)}{\partial x_i} (y_i - x_i) + \frac{1}{2!} \sum_{i,j=1}^{n} \frac{\partial^2 f(x)}{\partial x_i \, \partial x_j} (y_i - x_i)(y_j - x_j)$$

$$+ \cdots + \frac{1}{k!} \sum_{i_1, \cdots, i_k = 1}^{n} \frac{\partial^k f(x)}{\partial x_{i_1} \cdots \partial x_{i_k}} (y_{i_1} - x_{i_1}) \cdots (y_{i_k} - x_{i_k})$$

$$+ \frac{1}{(k+1)!} \sum_{i_1 \cdots i_{k+1} = 1}^{n} \frac{\partial^{k+1} f(\xi)}{\partial x_{i_1} \cdots \partial x_{i_{k+1}}} (y_{i_1} - x_{i_1}) \cdots (y_{i_{k+1}} - x_{i_{k+1}}).$$

We now obtain a variant of Theorem 2.

Theorem 3

If E is n-dimensional, $G \subset E$ is open, and

$$f: G \to R$$

has a continuous $(k + 1)$st differential, then if $x \in G$, $y \in G$, we have

$$f(y) = f(x) + \sum_{i=1}^{n} \frac{\partial f}{\partial x_i}(x)(y_i - x_i) + \cdots$$

$$+ \frac{1}{k!} \sum_{i_1, \cdots, i_k = 1}^{n} \frac{\partial^k f}{\partial x_{i_1} \cdots \partial x_{i_k}}(x) \cdot (y_{i_1} - x_{i_1}) \cdots (y_{i_k} - x_{i_k}) + R(x, y),$$

where

$$\lim_{|y - x| \to 0} \frac{|R(x, y)|}{|y - x|^k} = 0.$$

Remark In this theorem, $k + 1$ can be replaced by k, but the proof must be different. The present result is adequate for our needs.

Proof

By Theorem 2,

$$R(x, y) = \frac{1}{(k + 1)!} \sum_{i_1, \cdots, i_{k+1} = 1}^{n} \frac{\partial^{k+1} f}{\partial x_{i_1} \cdots \partial x_{i_{k+1}}}(\xi)$$

$$(y_{i_1} - x_{i_1}) \cdots (y_{i_{k+1}} - x_{i_{k+1}}),$$

whenever x and y lie on an open segment in G, and then ξ is between x and y on that segment.

Under the present hypothesis, there is an $r > 0$ and an M such that $|y - x| < r$ implies

$$\left| \frac{\partial^{k+1} f}{\partial x_{i_1} \cdots \partial x_{i_{k+1}}}(y) \right| < M,$$

for all choices of $i_1, \cdots, i_{k+1}$, so that

$$|R(x, y)| \leq Mn |y_{i_1} - x_{i_1}| \cdots |y_{i_{k+1}} - x_{i_{k+1}}|,$$

It follows that

$$\lim_{|y - x| \to 0} \frac{|R(x, y)|}{|y - x|^k} = 0. \quad \blacksquare$$

3. ABSOLUTE MAXIMA AND MINIMA

We consider maxima and minima for functions on subsets of E into R. In this section, we show that under certain conditions absolute maxima and minima exist.

Let E be n-dimensional, $S \subset E$, and

$$f: S \to R.$$

We say that f is **lower semi-continuous** at $x \in S$ if, for every $\epsilon > 0$, there is a $\delta > 0$ such that $y \in S$ and $|x - y| < \delta$ implies

$$f(y) > f(x) - \epsilon.$$

If f is lower semi-continuous at every $x \in S$, then it is said to be **lower semi-continuous** on S.

Theorem 4

If $K \subset E$ is compact and $f: K \to R$ is lower semi-continuous on K, then f has an absolute minimum on K, i.e., there is an $x^0 \in K$ such that $f(x^0) \leq f(x)$, for every $x \in K$.

Proof

We first note that the set

$$S = [f(x): x \in K]$$

has a lower bound. Suppose, for every n, there is an $x^n \in K$ such that $f(x^n) < -n$. Since K is compact, $\{x^n\}$ has a convergent subsequence which converges to an $x^0 \in K$. For every $\delta > 0$, there is an x^n with $|x^0 - x^n| < \delta$ and $f(x^n) < f(x^0) - 1$, contradicting the lower semi-continuity of f at x^0. Let

$$l = \inf S.$$

For every n, there is an $x^n \in K$ such that $f(x^n) < l + 1/n$. Now, $\{x^n\}$ has a convergent subsequence, $\{y^n\}$, which converges to an $x^0 \in K$. Now,

$$f(x^0) \leq \lim_{n \to \infty} f(y^n) = l,$$

so that $f(x^0) = l$. ∎

4. LOCATION OF MAXIMA AND MINIMA

The above theorem merely assures that a minimum exists but says nothing about its location. In this section, we consider the question of locating the maxima and minima, at least locally. First, we prove a theorem on interchanging the order in partial differentiation, and also say something about quadratic forms.

Theorem 5

If $f: G \rightarrow R$ is such that D^2f exists and is continuous, then for every $x^0 \in G$,

$$\frac{\partial^2 f}{\partial x_i \, \partial x_j}(x_0) = \frac{\partial^2 f}{\partial x_j \, \partial x_i}(x_0),$$

for every $i, j = 1, \cdots, n$.

Proof

It suffices to suppose E is two-dimensional and has coordinates (x, y). Let $(x_0, y_0) \in G$ and suppose $h > 0$, $k > 0$ are such that the rectangle $[x_0, x_0 + h] \times [y_0, y_0 + k] \subset G$.

We consider the second difference,

$$\frac{1}{hk}[\{f(x_0 + h, y_0 + k) - f(x_0, y_0 + k)\} - \{f(x_0 + h, y_0) - f(x_0, y_0)\}].$$

If we apply the mean-value theorem to the function

$$f(x, y_0 + k) - f(x, y_0),$$

this second difference is observed to be equal to

$$\frac{1}{k}\left[\frac{\partial f}{\partial x}(\xi, y_0 + k) - \frac{\partial f}{\partial x}(\xi, y_0)\right],$$

where $x_0 < \xi < x_0 + h$.

Now, by the mean-value theorem, this last expression is equal to

$$\frac{\partial^2 f}{\partial y \, \partial x}(\xi, \eta),$$

where $y_0 < \eta < y_0 + k$.

The above second difference thus converges to $\dfrac{\partial^2 f}{(\partial y \, \partial x)}(x_0, y_0)$ as h and k converge to zero.

By rearranging the second difference as

$$\frac{1}{hk}[\{f(x_0 + h, y_0 + k) - f(x_0 + h, y_0)\} - \{f(x_0, y_0 + k) - f(x_0, y_0)\}],$$

it also converges to $\dfrac{\partial^2 f}{(\partial x \, \partial y)}(x_0, y_0)$. ∎

Let E be an n-dimensional vector space. A mapping $f\colon E \times E \to R$ is called a **symmetric bilinear form** if

(a) $$f(x, y) = f(y, x),$$

(b) $$f(ax, y) = af(x, y),$$

(c) $$f(x^1 + x^2, y) = f(x^1, y) + f(x^2, y).$$

If f is a symmetric bilinear form we may associate with it a mapping

$$g\colon E \to R$$

by $g(x) = f(x, x)$. Such a mapping is called a **quadratic form**. In terms of coordinates in E, it is easily seen that a symmetric bilinear form has the representation

$$f(x, y) = \sum_{i,j=1}^{n} a_{ij} x_i y_j,$$

where the a_{ij} are real numbers with $a_{ij} = a_{ji}$ and where $x = (x_1, \cdots, x_n)$, $y = (y_1, \cdots, y_n)$. The corresponding quadratic form has the representation

$$g(x) = \sum_{i,j=1}^{n} a_{ij} x_i x_j.$$

We leave the proof to the reader.

A quadratic form is called **positive definite** if, for every vector $\xi = (\xi_1, \cdots, \xi_n) \neq (0, \cdots, 0)$, we have

$$\sum_{i,j=1}^{n} a_{ij} \xi_i \xi_j > 0.$$

It is called **negative definite** if, for every $\xi = (\xi_1, \cdots, \xi_n) \neq (0, \cdots, 0)$, we have

$$\sum_{i,j=1}^{n} a_{ij} \xi_i \xi_j < 0.$$

We now return to our problem.

If $G \subset E$ is open and $f\colon G \to R$, then $x \in G$ is said to be a **relative minimum** of f, if there is a $\delta > 0$ such that $|x - y| < \delta$ implies $f(x) \leq f(y)$. A **relative maximum** is defined similarly.

We shall consider only functions f which are continuously differentiable and shall first find a necessary condition for a relative minimum.

Proposition 1

If x^0 is a relative minimum for f, then the differential of f at x^0 is zero.

Proof

Suppose the differential at x^0 is not zero. Then there is an $i = 1, \cdots, n$ such that $\partial f/\partial x_i(x^0) \neq 0$. Suppose $\partial f/\partial x_i(x^0) > 0$. Then, for every $h > 0$, there is a $0 < \delta < h$ such that

$$f(x_1^0, \cdots, x_{i-1}^0, x_i^0 - \delta, x_{i+1}^0, \cdots, x_n^0) > f(x_1^0, \cdots, x_n^0). \quad \blacksquare$$

Evidently this condition is also necessary for a relative maximum.

However, it is possible for the differential of f to be zero at x^0, and for x^0 to be neither a relative maximum nor a relative minimum. For example, consider the function given by $f(x, y) = x^2 - y^2$. Then

$$\frac{\partial f}{\partial x}(x, y) = 2x$$

and

$$\frac{\partial f}{\partial y}(x, y) = 2y,$$

and the differential of f at the point $(0, 0)$ is zero. Clearly $f(0, 0) = 0$, $f(x, 0) > 0$, for all $x \neq 0$; and $f(0, y) < 0$, for all $y \neq 0$. Thus, $(0, 0)$ is a critical point which is neither a relative maximum nor a relative minimum.

The problem of finding sufficiency conditions for a critical point to be a relative maximum or relative minimum is more interesting. We have

Theorem 6

If $f: G \to R$ is of class C^2, where $G \subset E$ is open, and if $x^0 \in G$ is a critical point of f, then if the quadratic form

$$\sum_{i,j=i}^{n} \frac{\partial^2 f}{\partial x_i \partial x_j}(x^0) \cdot \xi_i \xi_j$$

is positive definite, the point x^0 is a relative minimum of f; and if the quadratic form is negative definite, the point x^0 is a relative maximum.

Proof

Suppose the quadratic form is positive definite. There is an $m > 0$ such that $|\xi| = 1$ implies

$$\sum_{i,j=1}^{n} \frac{\partial^2 f}{\partial x_i \, \partial x_j} (x^0) \xi_i \xi_j > 2m.$$

Let

$$M = \max \left[\sum_{i,j=1}^{n} |\xi_i \xi_j| : |\xi| = 1 \right]$$

and

$$\epsilon = m/M.$$

There is a $\delta > 0$ such that $|y - x^0| < \delta$ implies

$$\left| \frac{\partial^2 f}{\partial x_i \, \partial x_j} (y) - \frac{\partial^2 f}{\partial x_i \, \partial x_j} (x^0) \right| < \epsilon,$$

for every $i, j = 1, \cdots, n$.

Then $|y - x^0| < \delta$ and $|\xi| = 1$ implies

$$\left| \sum_{i,j=1}^{n} \frac{\partial^2 f}{\partial x_i \, \partial x_j} (y) \xi_i \xi_j - \sum_{i,j=1}^{n} \frac{\partial^2 f}{\partial x_i \, \partial x_j} (x^0) \xi_i \xi_j \right| < \epsilon \cdot M = m.$$

Hence, $|y - x^0| < \delta$ and $|\xi| = 1$ implies

$$\sum_{i,j=1}^{n} \frac{\partial^2 f}{\partial x_i \, \partial x_j} (y) \xi_i \xi_j > m.$$

But this implies that

$$\sum_{i,j=1}^{n} \frac{\partial^2 f}{\partial x_i \, \partial x_j} (y) \xi_i \xi_j > 0,$$

whenever $|\xi| > 0$.

By Theorem 2, $|y - x^0| < \delta$ implies there is an x, with $|x - x^0| < \delta$, and

$$f(y) = f(x^0) + \sum_{i=1}^{n} \frac{\partial f(x^0)}{\partial x_i} (y_i - x_i{}^0)$$

$$+ \frac{1}{2!} \sum_{i,j=1}^{n} \frac{\partial^2 f}{\partial x_i \, \partial x_j} (x)(y_i - x_i{}^0)(y_j - x_j{}^0).$$

Since x^0 is a critical point,

$$f(y) - f(x^0) = \frac{1}{2!} \sum_{i,j=1}^{n} \frac{\partial^2 f}{\partial x_i \, \partial x_j} (x)(y_i - x_i^0)(y_j - y_i^0) > 0,$$

so that $|y - x^0| < \delta$ implies $f(y) > f(x^0)$ and x^0 is a relative minimum. ∎

5. EXAMPLES

We make the criterion of Theorem 6 more explicit in the case where E is two-dimensional. Let the coordinates in E be (x, y). The quadratic form at issue is now

$$\frac{\partial^2 f}{\partial x^2}(x_0, y_0)\xi^2 + 2\frac{\partial^2 f}{\partial x\,\partial y}(x_0, y_0)\xi\eta + \frac{\partial^2 f}{\partial y^2}(x_0, y_0)\eta^2.$$

We now have

Theorem 7

If $G \subset R^2$ is open and $f\colon G \to R$ is of class C^2, and if $(x_0, y_0) \in G$ is a critical point of f, then

(a) if

$$\frac{\partial^2 f}{\partial x^2}(x_0, y_0) > 0$$

and

$$\frac{\partial^2 f}{\partial x^2}(x_0, y_0)\frac{\partial^2 f}{\partial y^2}(x_0, y_0) - \left(\frac{\partial^2 f}{\partial x\,\partial y}(x_0, y_0)\right)^2 > 0,$$

if follows that (x_0, y_0) is a relative minimum of f;

(b) if

$$\frac{\partial^2 f}{\partial x^2}(x_0, y_0) < 0$$

and

$$\frac{\partial^2 f}{\partial x^2}(x_0, y_0)\frac{\partial^2 f}{\partial y^2}(x_0, y_0) - \left(\frac{\partial^2 f}{\partial x\,\partial y}(x_0, y_0)\right)^2 > 0,$$

it follows that (x_0, y_0) is a relative maximum of f;

(c) if

$$\frac{\partial^2 f}{\partial x^2}(x_0, y_0)\frac{\partial^2 f}{\partial y^2}(x_0, y_0) - \left(\frac{\partial^2 f}{\partial x\,\partial y}(x_0, y_0)\right)^2 < 0,$$

then (x_0, y_0) is neither a relative minimum nor a relative maximum of f.

Proof

We need only study a quadratic form

$$ax^2 + 2bxy + cy^2.$$

If $a = c = 0$ this form is clearly not positive definite.

Suppose $a \neq 0$. In order for the form to be positive definite, we must have $a > 0$. Then

$$ax^2 + 2bxy + cy^2 = \frac{1}{a}(a^2x^2 + 2abxy + acy^2) = \frac{1}{a}[(ax + by)^2$$
$$+ (ac - b^2)y^2],$$

and it follows that we must have $ac - b^2 > 0$.

It is clear that if $a > 0$ and $ac - b^2 > 0$, the form is positive definite. Thus, this condition is necessary and sufficient, and implies $c > 0$. This proves part (**a**). The proof of part (**b**) is similar.

For part (**c**), it follows that the quadratic form is sometimes positive and sometimes negative on a set $E_n = [y: |x^0 - y| = h]$, and the result follows by an argument similar to the one used in proving Theorem 6. The details are left to the reader. ∎

We consider two examples.

Example A Let f be given by

$$f(x, y) = 2x^4 + y^4 - 2x^2 - 2y^2.$$

Then

$$\frac{\partial f}{\partial x}(x, y) = 8x^3 - 4x, \qquad \frac{\partial f}{\partial y}(x, y) = 4y^3 - 4y,$$

and there are nine critical points at

$$(0, 0), (0, 1), (0, -1), \left(\frac{1}{\sqrt{2}}, 0\right), \left(\frac{1}{\sqrt{2}}, 1\right), \left(\frac{1}{\sqrt{2}}, -1\right), \left(-\frac{1}{\sqrt{2}}, 0\right),$$

$$\left(-\frac{1}{\sqrt{2}}, 1\right), \text{ and } \left(-\frac{1}{\sqrt{2}}, -1\right).$$

Now

$$\frac{\partial^2 f}{\partial x^2}(x, y) = 24x^2 - 4,$$

$$\frac{\partial^2 f}{\partial y^2}(x, y) = 12y^2 - 4,$$

and

$$\frac{\partial^2 f}{\partial x\, \partial y}(x, y) = 0.$$

We then have the table

	(0, 0)	(0, 1)	(0,−1)	$\left(\frac{1}{\sqrt{2}},0\right)$	$\left(\frac{1}{\sqrt{2}},1\right)$	$\left(\frac{1}{\sqrt{2}},-1\right)$	$\left(-\frac{1}{\sqrt{2}},0\right)$	$\left(-\frac{1}{\sqrt{2}},1\right)$	$\left(-\frac{1}{\sqrt{2}},-1\right)$
$a = \dfrac{\partial^2 f}{\partial x^2}$	−4	−4	−4	8	8	8	8	8	8
$c = \dfrac{\partial^2 f}{\partial y^2}$	−4	8	8	−4	8	8	−4	8	8
$b = \dfrac{\partial^2 f}{\partial x\, \partial y}$	0	0	0	0	0	0	0	0	0
$ac - b^2$	16	−32	−32	−32	64	64	−32	64	64

In accordance with Theorem 7, $(0, 0)$ is a relative maximum,

$$\left(\frac{1}{\sqrt{2}}, 1\right), \left(\frac{1}{\sqrt{2}}, -1\right), \left(-\frac{1}{\sqrt{2}}, 1\right), \left(-\frac{1}{\sqrt{2}}, -1\right),$$

are relative minima, and

$$(0, 1), (0, -1), \left(\frac{1}{\sqrt{2}}, 0\right), \left(-\frac{1}{\sqrt{2}}, 0\right),$$

are neither relative minima nor relative maxima.

Our criterion thus gives a complete answer in this case.

Example B. Let f be given by

$$f(x, y) = x^2 - 3x^2 y + y^3.$$

Then

$$\frac{\partial f}{\partial x}(x, y) = 2x - 6xy$$

and

$$\frac{\partial f}{\partial y}(x, y) = -3x^2 + 3y^2,$$

so that the critical points are at

$$(0, 0), (\tfrac{1}{3}, \tfrac{1}{3}), \text{ and } (-\tfrac{1}{3}, \tfrac{1}{3}).$$

Now,

$$\frac{\partial^2 f}{\partial x^2}(x, y) = 2 - 6y,$$

$$\frac{\partial^2 f}{\partial y^2}(x, y) = 6y,$$

and

$$\frac{\partial^2 f}{\partial x\, \partial y}(x, y) = -6x.$$

We then have the table

	$(0, 0)$	$(\frac{1}{3}, \frac{1}{3})$	$(-\frac{1}{3}, \frac{1}{3})$
$a = \dfrac{\partial^2 f}{\partial x^2}$	2	0	0
$c = \dfrac{\partial^2 f}{\partial y^2}$	0	2	2
$b = \dfrac{\partial^2 f}{\partial x\, \partial y}$	0	-2	-2
$ac - b^2$	0	-4	-4

In accordance with Theorem 7, $(1/3, 1/3)$ and $(-1/3, 1/3)$ are neither relative maxima nor relative minima. Theorem 7 gives no information regarding the nature of the critical point $(0, 0)$. However, if we set $x = 0$, then $f(0, y) = y^3$, and it is clear that it is neither a relative maximum nor a relative minimum.

6. VOLUME OF A SET

The rest of this chapter is devoted to integration.

Let E be euclidean n-space and let there be an orthonormal basis in E with corresponding coordinates $(x_1, \cdots, x_n)$. We recall that a closed interval I is given by

$$I = [x: a_i \leq x_i \leq b_i, a_i < b_i, \quad i = 1, \cdots, n].$$

By the n-**dimensional volume** $v(I)$ of I, we mean the number

$$v(I) = \prod_{i=1}^{n} (b_i - a_i).$$

We shall also define sets of volume zero and the volume of open sets·

A set $S \subset E$ is said to be of **volume zero**, if for every $\epsilon > 0$, there is a finite set $I_1, I_2, \cdots, I_k$ of intervals such that

$$S \subset \bigcup_{j=1}^{k} I_j \quad \text{and} \quad \sum_{j=1}^{k} v(I_j) < \epsilon.$$

We note that if S has volume zero, then so does its closure $\bar{S}$. This is true since

$$S \subset \bigcup_{j=1}^{k} I_j$$

implies

$$\bar{S} \subset \bigcup_{j=1}^{k} I_j.$$

We now define the volume of an open set. Let $G \subset E$ be open. Then there is a sequence,

$$I_1, I_2, \cdots, I_j, \cdots,$$

of nonoverlapping closed intervals such that

$$G = \bigcup_{j=1}^{\infty} I_j.$$

It was implicit in the proof given in Chapter I of this fact, that this sequence may be chosen so that, for every compact $K \subset G$, there is a k such that

$$K \subset \bigcup_{j=1}^{k} I_j.$$

We now define the **volume** of G as

$$v(G) = \sum_{j=1}^{\infty} v(I_j),$$

where $I_1, I_2, \cdots$ is a sequence of nonoverlapping closed intervals in G such that

$$G = \bigcup_{j=1}^{\infty} I_j,$$

and every $K \subset G$ is contained in the union of finitely many I_j.

We note that this definition is independent of the sequence $I_1, I_2, \cdots$. For, let $J_1, J_2, \cdots$ be another such sequence. For every k, the set

$$\bigcup_{j=1}^{k} I_j$$

is contained in a set

$$\bigcup_{i=1}^{n(k)} J_i.$$

Hence

$$\sum_{j=1}^{k} v(I_j) \le \sum_{i=1}^{n(k)} v(J_i) \le \sum_{i=1}^{\infty} v(J_i).$$

The opposite inequality is similarly valid.

7.　INTEGRAL ON A CLOSED INTERVAL

Let $I \subset E$ be a closed interval and let $f: I \to R$ be a bounded real function. We shall define the upper and lower integrals of f. We say f is integrable if the upper and lower integrals are equal, and the common value is called the integral of f.

A **partition** $\pi = \{I_1, \cdots, I_k\}$ of a closed interval I is a finite set $I_1, \cdots, I_k$ of nonoverlapping closed intervals whose union is I.

The **diameter** of a set S is

$$d(S) = \sup\,[|x - y|: x \in S, y \in S].$$

The **norm** of a partition $\pi = \{I_1, \cdots, I_k\}$ of a closed interval I is the number

$$\|\pi\| = \max\,[\text{diameter } I_j: j = 1, \cdots, k].$$

A partition $\pi' = \{J_1, \cdots, J_m\}$ is said to be a **refinement** of $\pi = \{I_1, \cdots, I_k\}$, if every I_j is the union of intervals $J_{i_1}, \cdots, J_{i_r}$. We then write $\pi < \pi'$ or $\pi' > \pi$.

For every partition $\pi = \{I_1, \cdots, I_k\}$ of I, let

$$u(f, \pi) = \sum_{j=1}^{k} \sup\,[f(x): x \in I_j] \cdot v(I_j)$$

and

$$l(f, \pi) = \sum_{j=1}^{k} \inf\,[f(x): x \in I_j] \cdot v(I_j).$$

It is clear that, for every π, $u(f, \pi) \geq l(f, \pi)$. It is not quite as obvious, but easy enough for the details to be left to the reader, to show that, if $\pi' > \pi$, then

$$u(f, \pi') \leq u(f, \pi) \qquad \text{and} \qquad l(f, \pi') \geq l(f, \pi).$$

From these simple facts, we obtain the important result that, if π and π' are any partitions of I, then

$$u(f, \pi) \geq l(f, \pi').$$

In order to see this, let π'' be a partition of I such that $\pi'' > \pi$ and $\pi'' > \pi'$. Then

$$u(f, \pi) \geq u(f, \pi'') \geq l(f, \pi'') \geq l(f, \pi').$$

We now define the **upper integral** of f as

$$\overline{\int} f = \inf\,[u(f, \pi): \pi]$$

and the **lower integral** of f as

$$\underline{\int} f = \sup\,[l(f, \pi): \pi].$$

We have proved

Theorem 8

For every bounded f,

$$\overline{\int} f \geq \underline{\int} f,$$

and equality holds if and only if, for every $\epsilon > 0$, there is a partition π of I such that $u(f, \pi) - l(f, \pi) < \epsilon$.

If

$$\overline{\int} f = \underline{\int} f$$

we say that f is **integrable** on I, and write

$$\int f \quad \text{or} \quad \int_I f(x)\, dx.$$

Theorem 9

If $f: I \to R$ is continuous, then it is integrable.

Proof

Let $\epsilon > 0$. There is a $\delta > 0$ such that $x, y \in I$ and $|x - y| < \delta$ implies

$$|f(y) - f(x)| < \frac{\epsilon}{v(I)}.$$

Let $\pi = \{I_1, \cdots, I_k\}$ be a partition of I with $\|\pi\| < \delta$. Then

$$u(f, \pi) - l(f, \pi) = \sum_{j=1}^{k} \max\,[f(x): x \in I_j] \cdot v(I_j) - \sum_{j=1}^{k} \min[f(x): x \in I_j] \cdot v(I_j)$$

$$= \sum_{j=1}^{k} \{\max\,[f(x): x \in I_j] - \min\,[f(x): x \in I_j]\} \cdot v(I_j)$$

$$< \frac{\epsilon}{v(I)} \sum_{j=1}^{k} v(I_j) = \epsilon.$$

This proves the theorem. ∎

Theorem 10

If $f: I \rightarrow R$ and $\{I_1, \cdots, I_k\}$ is a partition of I, then f is integrable on I if and only if it is integrable on each $I_j, j = 1, \cdots, k$. Moreover,

$$\int_I f = \sum_{j=1}^k \int_{I_j} f.$$

The proof is left to the reader.

Theorem 11

If f and g are integrable on I, then $f + g$ is integrable and

$$\int f + g = \int f + \int g.$$

Proof

Let π be a partition of I. It is clear that

$$u(f + g, \pi) \leq u(f, \pi) + u(g, \pi)$$

and

$$l(f + g, \pi) \geq l(f, \pi) + l(g, \pi).$$

Let $\epsilon > 0$ and let π be such that

$$u(f, \pi) - l(f, \pi) < \frac{\epsilon}{2} \quad \text{and} \quad u(g, \pi) - l(g, \pi) < \frac{\epsilon}{2}.$$

Then $u(f + g, \pi) - l(f + g, \pi) < \epsilon$, so that f is integrable and

$$\left| \int f + g - \int f - \int g \right| \leq \epsilon. \quad \blacksquare$$

The proof of the next theorem will be simplified if we have a lemma whose proof is simple and will be omitted.

Lemma 1

If I is a closed interval and π is a partition of I, then for every $\epsilon > 0$, there is a $\delta > 0$ such that, for every partition π' of norm less than δ, the sum of the volumes of those intervals of the partition π', which are not contained in intervals of the partition π, is less than ϵ.

Theorem 12

If $f: I \rightarrow R$ is integrable, then for every $\epsilon > 0$, there is a $\delta > 0$ such that if $\pi = \{I_1, \cdots, I_k\}$ is a partition of I of norm less than δ, and if $x^j \in I_j, j = 1, \cdots, k$, then

$$\left| \int_I f - \sum_{j=1}^k f(x^j) v(I_j) \right| < \epsilon.$$

Proof

There is an M such that $|f(x)| \leq M$, for every $x \in I$. Let π be a partition of I such that $u(f, \pi) - l(f, \pi) < \epsilon/2$.

Let $\delta > 0$ be the δ of Lemma 1 corresponding to π and $\epsilon/2M$. Let π' be any partition of I of norm less than δ. Let $I_1, \cdots, I_r$ be those intervals in π' which are contained in intervals in π and let $I_{r+1}, \cdots, I_k$ be the remaining intervals in π'. Let $x^j \in I_j$, $j = 1, \cdots, k$. Then

$$\sum_{j=1}^{k} f(x^j)v(I_j) = \sum_{j=1}^{r} f(x^j)v(I_j) + \sum_{j=r+1}^{k} f(x^j)v(I_j) \leq u(f, \pi) + M \cdot \frac{\epsilon}{2M}$$

$$= u(f, \pi) + \frac{\epsilon}{2}.$$

Similarly,

$$\sum_{j=1}^{k} f(x^j)v(I_j) \geq l(f, \pi) - \frac{\epsilon}{2}.$$

Thus,

$$\int f - \epsilon \leq \sum_{j=1}^{k} f(x^j)v(I_j) \leq \int f + \epsilon. \quad \blacksquare$$

8. CONDITION FOR INTEGRABILITY

In this section, we still consider bounded functions on a closed interval I. We give necessary and sufficient conditions for such a function to be integrable.

For this purpose, we define the oscillation of a function f at a point x. For every $r > 0$, let

$$w_r(x) = \sup [f(y): |x - y| \leq r] - \inf [f(y): |x - y| \leq r].$$

The **oscillation** of f at x is then

$$w(x) = \lim_{r \to 0} w_r(x).$$

Clearly, the oscillation is nonnegative, and is zero at x if and only if f is continuous at x.

Theorem 13

If $f: I \to R$ is bounded, it is integrable if and only if, for every $k > 0$, the set of points at which the oscillation of f is greater than or equal to k has volume zero.

Proof

Suppose the condition holds. Let $\epsilon > 0$ and let

$$E_\epsilon = [x: w(x) \geq \epsilon].$$

Then E_ϵ can be covered by nonoverlapping closed intervals in I, the sum of whose volumes is less than $\epsilon/2M$, where $M = \sup [|f(x)|: x \in I]$. These intervals can be expanded to open intervals $J_1, \cdots, J_r$, the sum of whose volumes is still less than $\epsilon/2M$. By the same application of the Borel covering theorem as in the proof that a continuous function on a compact set is uniformly continuous, we may obtain a partition

$$I_1, \cdots, I_s \qquad \text{of} \qquad I - \bigcup_{i=1}^{r} J_i$$

such that, for each $j = 1, \cdots, s$ and $x, y \in I_j$, we have $|f(x) - f(y)| < \epsilon$. Form any partition π of I as $I_1, \cdots, I_s, I_{s+1}, \cdots, I_t$. Then

$$u(f, \pi) - l(f, \pi) \leq \epsilon v(I) + 2M \cdot \frac{\epsilon}{2M} = \epsilon[1 + v(I)].$$

Thus f is integrable.

Suppose, conversely, that there is a $k > 0$ such that the set of points

$$E_k = [x: w(x) \geq k]$$

does not have volume zero. Then there is a $\delta > 0$ such that if $I_1, \cdots, I_r$ covers E_k,

$$\sum_{j=1}^{r} v(I_j) > \delta.$$

Let $\pi = \{I_1, \cdots, I_m\}$ be a partition of I and let $I_1, \cdots, I_r$ be those intervals of I which contain points of E_k in their interior. Then

$$\sum_{j=1}^{r} v(I_j) > \delta.$$

It follows that

$$u(f, \pi) - l(f, \pi) \geq k \sum_{j=1}^{r} v(I_j) > k\delta$$

and that f is not integrable. ∎

We now consider sets S which are of the form

$$S = \bigcup_{j=1}^{\infty} S_j,$$

where S_j has volume zero.

Lemma 2

   If S is compact and

$$S = \bigcup_{j=1}^{\infty} S_j, \qquad v(S_j) = 0, j = 1, 2, \cdots,$$

then $v(S) = 0$.

Proof

   Let $\epsilon > 0$. For each j, there is a finite set

$$I^{(j)}_1, \cdots, I^{(j)}_{n_j}$$

of open intervals with

$$S_j \subset \bigcup_{k=1}^{n_j} I^{(j)}_k$$

and

$$\sum_{k=1}^{n_j} v(I^{(j)}_k) < \frac{\epsilon}{2^j}.$$

The union of all these open intervals covers S. By the Borel covering theorem, a finite set of them, the sum of whose volumes is less than ϵ, covers S. Hence $v(S) = 0$. ∎

Theorem 14

   If $f: I \to R$ is bounded, then f is integrable if and only if its set D of points of discontinuity is the union of a countable set of sets of volume zero.

Proof

For every $n = 1, 2, \cdots$, let

$$D_n = \left[x: w(x) \geq \frac{1}{n} \right].$$

Then D_n is compact and

$$D = \bigcup_{n=1}^{\infty} D_n.$$

Suppose the condition of the theorem holds. Then, by Lemma 2, each D_n has volume zero, so that f is integrable by Theorem 13. Conversely if f is integrable, each D_n has volume zero, so that D has the required property. ∎

9. INTEGRAL ON AN OPEN SET

   Let $f: G \to R$, where $G \subset E$ is open. We suppose f is nonnegative and integrable on every closed interval in G. While this implies that f is bounded on every closed interval in G, f may be unbounded on G itself.

Let I_1, I_2, $\cdots$ be nonoverlapping closed intervals whose union is G, and such that every compact $K \subset G$ is contained in the union of a finite number of them. Then define the integral of f on G as

$$\int_G f = \sum_{k=1}^{\infty} \int_{I_k} f.$$

That the value of the integral is independent of the partition follows from Theorem 10 and the fact that the integral of a nonnegative function is nonnegative. The value may be infinite.

In order to define the integral of a function f which is not non-negative, we associate with every f the functions f^+ and f^- defined by

$$f^+(x) = \max{(f(x), 0)}$$

and

$$f^-(x) = -\min{(f(x), 0)}.$$

Then, $f = f^+ - f^-$, and f^+, f^- are nonnegative. We state the preliminary

Theorem 15

If I is a closed interval and $f: I \to R$, then f is integrable if and only if f^+ and f^- are integrable, and then

$$\int f = \int f^+ - \int f^-.$$

Proof

It is only necessary to show that the integrability of f implies that of f^+. The remaining statements follow immediately from what we have already done.

Suppose f is integrable on I. Let f be continuous at x. If $f(x) > 0$, there is a sphere $\sigma(x, r)$ on which f is positive, so that $f(y) = f^+(y)$, for every $y \in \sigma(x, r)$, and hence f^+ is continuous at x. If $f(x) < 0$, there is a sphere $\sigma(x, r)$ on which f is negative, so that $f^+(y) = 0$, for every $y \in \sigma(x, r)$, and hence f^+ is continuous at x. If $f(x) = 0$, given $\epsilon > 0$, there is a sphere $\sigma(x, r)$ on which $-\epsilon < f(y) < \epsilon$. Then $0 \leq f^+(y) < \epsilon$, $y \in \sigma(x, r)$, and f^+ is again continuous at x.

Thus, the set of points of discontinuity of f^+ is a subset of the set of points of discontinuity of f and, by Theorem 14, f^+ is integrable on I. ∎

Now, let $f: G \to R$, where $G \subset E$ is open. We say that f is **integrable** on G if it is integrable on every closed interval $I \subset G$ and if at least one of the integrals

$$\int_G f^+ \quad \text{and} \quad \int_G f^-$$

is finite. If

$$\int_G f^+ < \infty, \quad \int_G f^- < \infty,$$

we define

$$\int_G f$$

by

$$\int_G f = \int_G f^+ - \int_G f^-.$$

If

$$\int_G f^+ = \infty, \quad \int_G f^- < \infty,$$

we define

$$\int_G f = \infty,$$

and if

$$\int_G f^+ < \infty, \quad \int_G f^- = \infty,$$

we define

$$\int_G f = -\infty.$$

We may state

Theorem 16

If f is integrable on G, then for every nonoverlapping $I_1, I_2, \cdots$ whose union is G such that every compact $K \subset G$ is in a finite number of them, we have

$$\int_G f = \sum_{k=1}^{\infty} \int_{I_k} f.$$

The proof is left for the reader.

We define $f: G \to R$ to be **summable** if

$$\int_G |f| < \infty.$$

10. ITERATED INTEGRAL

In evaluating integrals, it is usually necessary to know that they are equal to iterated integrals. This is a very delicate process within our present framework, which becomes simple and quite general, using a more general definition of integral (the Lebesgue integral). Nevertheless, we can obtain results which have wide application.

We first give a counterexample. Let $I = [0, 1] \times [0, 1]$ and let $S \subset I$ be a dense set in I such that, for each $x \in [0, 1]$, there is at most one y, with $(x, y) \in S$, and for each $y \in [0, 1]$, there is at most one x, with $(x, y) \in S$. We leave it to the reader to construct such an example and to show that the function $f: I \to R$ which is 1 on S and 0 on $I - S$ is not integrable. However, for each $x_0 \in [0, 1]$, the function $f(x_0, y)$ is integrable in y and

$$\phi(x_0) = \int f(x_0, y)\, dy = 0.$$

Thus, ϕ is an integrable function of x, and its integral exists and is zero. Thus,

$$\int dx \int f(x, y)\, dy = 0.$$

Similarly,

$$\int dy \int f(x, y)\, dx = 0.$$

We thus have a function whose iterated integrals exist, but the function itself is not integrable.

On the other hand, the next theorem shows that if f is integrable, then the iterated integrals exist and yield the value of the integral.

Theorem 17

If $I \subset E$ is a closed interval, E is n-dimensional, and $f: I \to R$ is integrable; and if $I = [a, b] \times J$, where J is an $(n - 1)$-dimensional interval, then for every $x \in [a, b]$, except possibly for those belonging to a set of one-dimensional volume zero, $f(x, y)$ is an integrable function of y, and

$$\int_I f = \int_a^b dx \int_J f(x, y)\, dy.$$

Proof

Let $\epsilon > 0$. Consider a partition π of I, obtained from a partition π' of J and a partition π'' of $[a, b]$, where the intervals of π are the

$[a_j, b_j] \times J_i$, where $[a_j, b_j]$ is an interval of π'' and J_i is an interval of π', such that

$$\left| \int_I f - \sum f(\xi_k) v(I_k) \right| < \epsilon,$$

and where $\xi_k \in I_k$ is arbitrary.

Let

$$\phi(x) = \overline{\int} f(x, y)\, dy$$

and

$$\psi(x) = \underline{\int} f(x, y)\, dy,$$

for every $x \in [a, b]$. Then, by the choice of π, we readily have

$$\left| \int_I f - \overline{\int} \phi(x)\, dx \right| < \epsilon$$

and

$$\left| \int_I f - \underline{\int} \psi(x)\, dx \right| < \epsilon.$$

Since this holds for every $\epsilon > 0$, we obtain

$$\int_I f = \overline{\int} \phi(x)\, dx = \underline{\int} \psi(x)\, dx.$$

Since $\phi(x) \geq \psi(x)$, $x \in [a, b]$, this implies $\phi(x) = \psi(x)$, except perhaps on a set of one-dimensional volume zero. Then

$$\int f(x, y)\, dy$$

exists and

$$\int dx \int f(x, y)\, dy = \int f. \quad \blacksquare$$

We now state the more general

Theorem 18

If $G \subset E$ is open and $\bar{G}$ is the closure of G; if for every $\epsilon > 0$, there is a finite set, $I_1, \cdots, I_k$, of nonoverlapping closed intervals contained in G, and a finite set, $J_1, \cdots, J_r$, of nonoverlapping closed intervals containing $\bar{G}$, such that

$$\sum_{i=1}^{r} v(J_i) < \sum_{j=1}^{k} v(I_j) + \epsilon,$$

and if for every real x,

$$G_x = \{y : (x, y) \in G\},$$

then for every bounded integrable f on G,

$$\int_G f = \int dx \int_{G_x} f(x, y) \, dy.$$

The proof follows without great difficulty from Theorem 17. We leave the details to the reader.

11.   VOLUME OF n-BALL

As an example, we find a formula for the n-dimensional volume of an n-dimensional ball of radius r; i.e., the set

$$[x: |x| < r].$$

The one-dimensional ball of radius r is the set of real numbers $-r < x < r$, and its volume is

$$v_1(r) = 2r.$$

We find the formula for the volume of the n-dimensional ball of radius r by induction. Suppose

$$v_{n-1}(r) = a_{n-1} r^{n-1},$$

where a_{n-1} is a constant.

Let G be the n-ball of radius r. For every real x, with $-r < x < r$, the section G_x is an $(n-1)$-ball of radius $(r^2 - x^2)^{1/2}$. Moreover, the conditions of Theorem 18 are satisfied by G and the constant function 1, so that the volume of G is given by

$$v_n(r) = \int_G 1 = \int_{-r}^{r} dx \int_{G_x} 1.$$

But

$$\int_{G_x} 1$$

is the volume of the $(n-1)$-ball of radius $(r^2 - x^2)^{1/2}$, so that it equals

$$a_{n-1}(r^2 - x^2)^{(n-1)/2}.$$

Thus

$$v_n(r) = a_{n-1} \int_{-r}^{r} (r^2 - x^2)^{(n-1)/2}\, dx = 2a_{n-1}r^n \int_{0}^{\pi/2} \cos^n \theta\, d\theta$$

$$= 2r \int_{0}^{\pi/2} \cos^n \theta \cdot d\theta \cdot v_{n-1}(r).$$

In particular, we have

$$v_1(r) = 2r$$
$$v_2(r) = \pi r^2$$
$$v_3(r) = \tfrac{4}{3}\pi r^3$$
$$v_4(r) = \tfrac{1}{2}\pi^2 r^4$$
$$v_5(r) = \tfrac{8}{15}\pi^2 r^5$$
$$\cdots$$

In making these computations, we may use Wallis's formula, which says

$$\int_{0}^{\pi/2} \cos^n \theta\, d\theta = \begin{cases} \dfrac{(n-1)(n-3)\cdots}{n \cdot (n-2) \cdots}, & \text{if } n \text{ is odd,} \\[2ex] \dfrac{(n-1)(n-3)\cdots}{n(n-2)\cdots} \cdot \dfrac{\pi}{2}, & \text{if } n \text{ is even.} \end{cases}$$

We also have the formula

$$v_n(r) = 2^n r^n \int_{0}^{\pi/2} \cos \theta\, d\theta \int_{0}^{\pi/2} \cos^2 \theta\, d\theta \cdots \int_{0}^{\pi/2} \cos^n \theta\, d\theta.$$

12. INTERCHANGE OF ORDER OF INTEGRATION WITH DIFFERENTIATION

In this section, we show that under certain conditions it is possible to interchange the operations of differentiation and integration.

We prove

Theorem 19

If $I = [a, b] \times [c, d]$ is a closed rectangle and if f and $\partial f/\partial y$ are continuous on I, then if

$$F(y) = \int_{a}^{b} f(x, y)\, dx,$$

it follows that F is differentiable and that

$$F'(y) = \int_a^b \frac{\partial f(x, y)}{\partial y} \, dx.$$

Proof

We note that

$$F(y + h) - F(y) = \int_a^b \{f(x, y + h) - f(x, y)\} \, dx = \int_a^b h \frac{\partial f}{\partial y}(x, \eta(x)) \, dx,$$

where $y \le \eta(x) \le y + h$, by the mean-value theorem. Then

$$\left| \frac{F(y + h) - F(y)}{h} - \int_a^b \frac{\partial f}{\partial y}(x, y) \, dx \right| \le \int_a^b \left| \frac{\partial f}{\partial y}(x, y) - \frac{\partial f}{\partial y}(x, \eta(x)) \right| \, dx.$$

and the theorem follows from the uniform continuity of $\partial f / \partial y$ on I. ∎

We state the more general

Theorem 20

Let f and $\partial f / \partial y$ be continuous on an open set G and let a closed region bounded by continuously differentiable $x = \phi(y)$, $x = \psi(y)$, and $y = c$ $y = d$, with $\phi(y) < \psi(y)$, for all $y \in [c, d]$, be contained in G. Then, if

$$F(y) = \int_{\phi(y)}^{\psi(y)} f(x, y) \, dx, \qquad y \in [c, d],$$

it follows that F is differentiable and that

$$F'(y) = \int_{\phi(y)}^{\psi(y)} \frac{\partial f}{\partial y}(x, y) \, dx + f(\psi(y), y)\psi'(y) - f(\phi(y), y)\phi'(y).$$

Proof

We note that

$$F(y + h) - F(y) = \int_{\phi(y)}^{\psi(y)} f(x, y + h) - \int_{\phi(y)}^{\psi(y)} f(x, y) \, dx$$

$$+ \int_{\psi(y)}^{\psi(y + h)} f(x, y + h) \, dx - \int_{\phi(y)}^{\phi(y + h)} f(x, y + h) \, dx$$

$$= \int_{\phi(y)}^{\psi(y)} h \frac{\partial f}{\partial y}(x, \eta(x)) \, dx$$

$$+ f(\xi, y + h)[\psi(y + h) - \psi(y)]$$

$$- f(\xi', y + h)[\phi(y + h) - \phi(y)],$$

where $y \leq \eta(x) \leq y + h$, for each x, $\psi(y) \leq \xi \leq \psi(y + h)$, and $\phi(y) \leq \xi' \leq \phi(y + h)$. But again by the mean-value theorem

$$F(y + h) - F(y) = \int_{\phi(y)}^{\psi(y)} h \frac{\partial f}{\partial y}(x, \eta(x)) \, dx$$

$$+ f(\xi, y + h)\psi'(u) \cdot h - f(\xi', y + h)\phi'(v) \cdot h,$$

where $y < u < y + h$ and $y < v < y + h$.
  Then

$$\left| \frac{F(y + h) - F(y)}{h} - \left\{ \int_{\phi(y)}^{\psi(y)} \frac{\partial f}{\partial y}(x, y) \, dx + f(\psi(y), y)\psi'(y) \right. \right.$$

$$\left. \left. - f(\phi(y), y)\phi'(y) \right\} \right|$$

$$\leq \int_{\phi(y)}^{\psi(y)} \left| \frac{\partial f}{\partial y}(x, y) - \frac{\partial f}{\partial y}(x, \eta(x)) \right| dx$$

$$+ M\{|\psi'(u) - \psi'(y)| + |\phi'(v) - \phi'(y)|\},$$

where $M = \max [|f(x, y)|: (x, y)$ in our closed region].
  The theorem follows from the uniform continuity of $\partial f / \partial y$ and of ψ' and ϕ'. ∎

EXERCISES

1.1 Show that the series

$$1 - \frac{x^3}{3!} + \cdots + (-1)^n \frac{x^{2n+1}}{(2n + 1)!} + \cdots$$

converges to $\sin x$, for every real x, and the convergence is uniform on every compact set of reals.

1.2 Show, however, that the convergence is not uniform on the set of all reals.

1.3 For $0 < x < \pi/2$, show that

$$\frac{2}{\pi} < \frac{\sin x}{x} < 1.$$

1.4 Suppose f is of class C^{n+1} in an open interval I, with $x_0 \in I$. Show that for every $x \in I$,

$$f(x) = f(x_0) + f'(x_0)(x - x_0) + \cdots + \frac{f^n(x_0)}{n!}(x - x_0)^n$$

$$+ \frac{1}{n!}\int_0^x (x - t)^n f^{n+1}(t)\, dt.$$

1.5 Suppose f is continuous in an open interval I, $x_0 \in I$, and $f^2(x_0)$ exists. Show that

$$f^2(x_0) = \lim_{h \to 0} \frac{f(x_0 + h) + f(x_0 - h) - 2f(x_0)}{h^2}.$$

1.6 Let $a_0 + a_1 x + \cdots + a_n x^n + \cdots$ be a power series in the real variable x. Show that there is an r such that the series converges for $|x| < r$ and diverges for $|x| > r$. (r may be 0 or $+\infty$.)

1.7 Give an expression for the value of r in terms of the coefficients $a_0, a_1, \cdots, a_n, \cdots$.

2.1 If f is of class C^∞, show that

$$f(x) = f(x^0) + \sum_{i=1}^n \frac{\partial f}{\partial x_i}(x^0)(x_i - x_i^0) + \sum_{i,j=1}^n g_{ij}(x)(x_i - x_i^0)(x_j - x_j^0),$$

where the g_{ij} are of class C^∞.

2.2 Given a series

$$a_0 + a_1 x + a_2 y + \sum_{i,j=1}^n a_{ij} x^i y^j + \cdots,$$

discuss as much as you can the nature of the set of (x, y) for which the series converges.

3.1 If f is continuous on $[0, 1]$, show that for every n, there is a polynomial of degree n such that

$$\max |f(x) - p(x)| \le \max |f(x) - q(x)|,$$

for every polynomial q of degree n.

3.2 Define upper semicontinuity. If $f: R^n \to R$ for an open sphere $\sigma \subset R^n$, the oscillation of f in σ is

$$\omega(f, \sigma) = \sup [f(x): x \in \sigma] - \inf [f(x): x \in \sigma].$$

The oscillation $\omega(f; x)$ of f at x is then

$$\inf [\omega(f; \sigma): \sigma \quad \text{such that } x \in \sigma].$$

Show that for every bounded f, the function $\omega(f; x)$ is upper semi-continuous.

4.1 If $f: R^2 \to R$ is such that, on an open set G,

$$\frac{\partial f}{\partial x}, \quad \frac{\partial f}{\partial y}, \quad \text{and} \quad \frac{\partial^2 f}{\partial x \partial y}$$

exist and are continuous, show that $\partial^2 f / \partial y \partial x$ exists and is continuous on G and that

$$\frac{\partial^2 f}{\partial x \partial y} = \frac{\partial^2 f}{\partial y \partial x}.$$

4.2 If $f: R^2 \to R$ is such that $\partial f / \partial x$, $\partial f / \partial y$ exist and are continuous in a neighborhood of (x_0, y_0), and the differentials of the functions $\partial f / \partial x$, $\partial f / \partial y$ exist at this point, show that

$$\frac{\partial^2 f}{\partial x \partial y}(x_0, y_0) = \frac{\partial^2 f}{\partial y \partial x}(x_0, y_0).$$

4.3 If f is a bilinear form, show that the corresponding quadratic form has the representation

$$g(x) = \sum_{i,j=1}^{n} a_{ij} x_i x_j.$$

5.1 Obtain an analogue of Theorem 7 for the three-dimensional case.

5.2 Discuss the critical points of

$$f(x, y) = x^3 - x^2 y + 3y^2.$$

5.3 Given n real numbers $x_1, \cdots, x_n$, find the number x for which

$$\sum_{i=1}^{n} (x - x_i)^2$$

is a minimum.

5.4 Extend to the case of k points in n-space.

5.5 Find the volume of the greatest interval which can be placed inside an ellipsoid.

5.6 Find the point in a plane such that the sum of the squares of its distances from the vertices of a triangle is smallest.

5.7 Answer the analogous question for a tetrahedron.

5.8 Find the maximum value of $x_1 \cdots x_n$, $x_i > 0$, $i = 1, \cdots, n$ subject to the condition that $x_1 + \cdots + x_n = n$.

5.9 Show that if $x_i > 0$, $i = 1, \cdots, n$, then

$$(x_1 \cdots x_n)^{1/n} \leq \frac{x_1 + \cdots + x_n}{n}.$$

5.10 Find the shortest distance from a point to a hyperplane in R^n.

5.11 A silo is in the form of a cylinder topped by a cone. What are the most economical dimensions?

6.1 Show that if X is a proper subspace of euclidean n-space, then every compact subset of X has volume zero.

6.2 Give an example of a subset of R^2, which is homeomorphic to the circle but whose two dimensional volume is not zero.

6.3 Show that the union of finitely many sets of volume zero is a set of volume zero.

7.1 Show that for every partition π,

$$u(f, \pi) \geq l(f, \pi).$$

7.2 Prove Theorem 10.

7.3 Prove Lemma 1.

9.1 Prove Theorem 16.

10.1 In the proof of Theorem 17, supply details as to why

$$\left| \int f - \overline{\int} \phi(x) \, dx \right| < \epsilon.$$

10.2 If f and g are integrable on [a,b] show that

$$\int_a^b |f - g| = 0$$

if and only if $f(x) = g(x)$, except on a set of one-dimensional volume zero.

10.3 Prove Theorem 18.

10.4 Show that, on a closed interval, the product of integrable functions is integrable.

10.5 Show, by example, that this is false for integrable functions on an open set.

10.6 Give an example of a dense set in the plane which has at most one point on each line parallel to either axis.

11.1 A 1 in. metal cube has 3, 1/2 in. (in diameter) holes bored with axes perpendicular to the faces at their centers. What is the volume of the remaining metal?

11.2 Show that the force of attraction of a solid sphere, of uniform density, for a point outside the sphere is the same as it would be if the mass of the sphere were concentrated at its center.

11.3 A square hole, of side a, is bored through a sphere of radius $b > a$. What is the volume remaining if the axis of the hole passes through the center of the sphere?

MAIN THEOREMS
ON MAPPINGS

In this chapter we define regular, continuously differentiable mappings on n-space into n-space. We show that such mappings are locally one-one with continuously differentiable inverse. We then use this theorem to prove the implicit function theorem. Next, we discuss determinants and oriented volumes. A transformation of integrals formula under change of variable is derived and applied to obtain some important probability density functions of statistics. Finally, a brief treatment is given of length of a curve and area of a surface.

1. REGULAR ELEMENTS IN L(E, F)

We present some preliminary facts about linear mappings in this section.

Let E be a vector space and let H be a subspace of E. We define the **quotient space** E/H. The elements of this space are subsets of E. They may be defined as the sets

$$H_x = x + H = [x + y : y \in H],$$

for all $x \in E$. It is easy to see that for every $x \in E$, $y \in E$, we have either

$$H_x = H_y \qquad \text{or} \qquad H_x \cap H_y = \emptyset.$$

Indeed, $H_x = H_y$ if and only if $x - y \in H$. The set E/H is thus well defined. We assert that it is a vector space with operations

$$H_x + H_y = H_{x+y}$$

and

$$aH_x = H_{ax}.$$

We leave the details to the reader.

Suppose

$$x_1, x_2, \cdots, x_k, x_{k+1}, \cdots, x_n$$

is a basis for E, with $x_1, \cdots, x_k$ a basis for H. It is then an easy matter to show that

$$H_{x_{k+1}}, \cdots, H_{x_n}$$

is a basis for E/H. We also leave this as an exercise for the reader. It follows from this that

$$\dim H + \dim E/H = \dim E.$$

Now, let E and F be vector spaces and let $f \in L(E, F)$. We recall that the kernel of f is the subspace $K \subset E$ which is taken by f into θ, and the image (or range) of f is the subspace $R \subset F$ of points which are maps $f(x)$ of points $x \in E$.

Since $f(x) = f(y)$ if and only if $f(x - y) = \theta$, i.e., $x - y \in K$, it follows that $f(x) = f(y)$ if and only if $K_x = K_y$. Thus E/K is isomorphic with R. We thus have the fact that for every $f \in L(E, F)$,

$$\dim K + \dim R = \dim E,$$

where K is the kernel of f and R is the image of f.

As a corollary, f is injective if and only if $\dim R = \dim E$.

Thus, if E and F have the same dimension, f is injective if and only if it is bijective, i.e., it has an inverse.

Proposition 1

If E and F have the same dimension, then $f \in L(E, F)$ is one-one if and only if it is an isomorphism.

Such an f will be called **regular**. Otherwise, f will be called **singular**. f is singular if and only if its kernel contains more than just θ, or its image is a proper subspace of F.

We suppose, in the next few sections, that E and F are both n-dimensional. Let E and F be normed. The following observations are pertinent to what follows:

(a) The set, $U \subset L(E, F)$, of regular linear mappings is open in $L(E, F)$.

Proof

Let $f \in U$. Then
$$f(x) \neq 0 \qquad \text{if } |x| = 1.$$
Since the set for which $|x| = 1$ is compact and since f is continuous, there is an $m > 0$ such that
$$|f(x)| \geq m, \qquad \text{whenever } |x| = 1.$$
By linearity,
$$|f(x)| \geq m\, |x|, \qquad \text{for all } x \in E.$$
Let $G \subset L(E, F)$ be the set of all g such that
$$|f - g| < \frac{m}{2}.$$
Then, for $g \in G$,
$$|g(x)| \geq \frac{m}{2}, \qquad \text{whenever } |x| = 1.$$
It follows that $g \in U$, so that U is open.   ∎

(b)   For every regular $f \in L(E, F)$, the inverse $f^{-1} \in L(F, E)$ is regular.

This is obvious from the definition of regular mappings.   ∎

Let U be the open set of regular mappings in $L(E, F)$ and let V be the open set of regular mappings in $L(F, E)$. We consider the mapping
$$\phi: U \to V$$
defined by $\phi(f) = f^{-1}$. It is evidently a bijective mapping. Moreover, we shall see that both ϕ and its inverse are continuous, i.e., ϕ is a homeomorphism.

(c)   The mapping
$$\phi: U \to V$$
is a homeomorphism.

Proof

We need only show that ϕ is continuous. For this, let f be regular, and note that there are δ, M, and m, all positive, such that $|f - g| < \delta$ implies $|g^{-1}| \leq M$, and $|g(x)| \geq m\, |x|$, for all $x \in E$. Let $0 < \epsilon < \delta$, and suppose $|f - g| < \epsilon$. Let $y \in F$. Then
$$m\, |f^{-1}(y) - g^{-1}(y)| \leq |f(f^{-1}(y) - g^{-1}(y))| = |y - f \circ g^{-1}(y)|$$
$$= |g \circ g^{-1}(y) - f \circ g^{-1}(y)| \leq |g - f|\, |g^{-1}|\, |y| < \epsilon M\, |y|.$$

It follows that

$$|f^{-1} - g^{-1}| < \frac{M}{m} \epsilon,$$

and ϕ is continuous. ∎

2. INVERSE OF A MAPPING

We again let E and F be n-dimensional vector spaces. Let $G \subset E$ be open and let

$$f: G \to F$$

be continuously differentiable on G.
We need

Lemma 1
We have

$$f(y) = f(x) + l_x(y - x) + R(x, y),$$

where l_x is the differential of f at $x \in G$ and where for every compact $K \subset G$,

$$\lim_{|x-y| \to 0} \frac{|R(x, y)|}{|x - y|} = 0,$$

where the convergence is uniform in x, $x \in K$.

Proof
For real functions of one variable, the mean-value theorem asserts that

$$f(y) = f(x) + f'(x)(y - x) + [f'(\xi) - f'(x)](y - x),$$

where ξ lies between x and y. Letting $R(x, y) = [f'(\xi) - f'(x)](y - x)$, the theorem follows from the uniform continuity of f'.

In the general case, we need only set coordinates in E and F and apply the above to each entry of the Jacobian matrix of f. We leave the details as an exercise. ∎

We say that a differentiable mapping is **regular** at x if its differential at x is regular. We may now prove

Theorem 1
If E and F are n-dimensional vector spaces, $G \subset E$ is open, and

$$f: G \to F$$

is continuously differentiable on G and regular at $x \in G$, then there is an open sphere $\sigma(x, r) \subset G$ such that

$$f: \sigma(x, r) \to F$$

is injective.

Proof

For every $y \in G$, let l_y be the differential of f at y. Since f is regular at x, there is an $m > 0$ such that

$$|l_x(z)| \geq 2m \, |z|,$$

for every $z \in E$. But f is continuously differentiable. Hence, there is a closed sphere $\bar{\sigma}(x, \rho) \subset G$ such that for every $y \in \bar{\sigma}(x, \rho)$, we have

$$|l_y(z)| \geq m \, |z|,$$

for every $z \in E$.

By the lemma, there is an $r > 0$ such that for every $y \in \bar{\sigma}(x, \rho)$ and $|z - y| < 2r$, we have $z \in G$ and

$$|f(z) - f(y) - l_y(z - y)| \leq \frac{m}{2} |z - y|$$

Suppose $r < \rho$, and consider the open sphere $\sigma(x, r)$. Let $y, z \in \sigma(x, r)$. Then

$$|l_y(z - y)| \geq m \, |z - y|,$$

so that, if $z \neq y$,

$$|f(z) - f(y)| \geq m \, |z - y| - \frac{m}{2} |z - y| > 0,$$

and $f(z) \neq f(y)$. ∎

As a complement to the above theorem, we have

Theorem 2

If E and F are n-dimensional vector spaces, $G \subset E$ is open, and

$$f: G \to F$$

is continuously differentiable and regular on G, then the image, $f(G)$, of G is open in F.

Proof

Let $x^0 \in G$. There is an open sphere $\sigma(x^0, \rho)$ such that

$$f: \sigma(x^0, \rho) \to F$$

is injective. Let $0 < r < \rho$ and let

$$S = [x: |x^0 - x| = r].$$

Then S is compact, so that its image $f(S)$ is compact. (The image of a compact set with respect to a continuous mapping is easily seen to be compact.) Now, $f(x^0) \notin f(S)$, so that the distance,

$$\min \left[|f(x^0) - f(x)| : x \in S \right] = m > 0,$$

of $f(S)$ from $f(x^0)$ is positive. Let $y \in F$ be such that

$$|f(x^0) - y| < \frac{m}{3}.$$

We endow E and F with orthonormal bases. In the corresponding coordinates, we consider the function

$$\phi(x) = [f_1(x_1, \cdots, x_n) - y_1]^2 + \cdots + [f_n(x_1, \ldots, x_n) - y_n]^2,$$

where $(x_1, \cdots, x_n)$ are the coordinates of x and $(y_1, \cdots, y_n)$ the coordinates of y. Then ϕ is the square of the distance from $f(x)$ to y.

This continuous function ϕ has an absolute minimum on the closed sphere $\bar{\sigma}(x^0, r)$. Since

$$|f(x^0) - y| < |f(x) - y|,$$

for every $x \in S$, this minimum is an interior point of $\bar{\sigma}(x^0, r)$, i.e., it is in $\sigma(x^0, r)$. It thus must be a critical point of ϕ. It must, accordingly, satisfy the conditions

$$\frac{\partial \phi}{\partial x_1}(x) = \cdots = \frac{\partial \phi}{\partial x_n}(x) = 0.$$

But these conditions say that

$$(f_1(x) - y_1)\frac{\partial f_1}{\partial x_1}(x) + \cdots + (f_n(x) - y_n)\frac{\partial f_n}{\partial x_1}(x) = 0$$

$$\cdots$$

$$(f_1(x) - y_1)\frac{\partial f_1}{\partial x_n}(x) + \cdots + (f_n(x) - y_n)\frac{\partial f_n}{\partial x_n}(x) = 0.$$

The regularity of f implies that the only vector taken into the zero vector by the differential of f is the zero vector. Thus

$$f_1(x) - y_1 = 0, \cdots, f_n(x) - y_n = 0,$$

and there is an $x \in \sigma(x^0, r)$ such that $f(x) = y$. In other words,

$$f(G) \supset \sigma\left(f(x^0), \frac{m}{3}\right),$$

and $f(G)$ is open. ∎

As the final result of this section, we obtain a companion to the above two results.

Theorem 3

If E and F are n-dimensional vector spaces, $G \subset E$ is open,

$$f: G \to F$$

is continuously differentiable and regular at every point of G, and f is injective, then f^{-1} is continuously differentiable and regular on $f(G)$.

Proof

We first note that if l_x is the differential of f at $x \in G$, then l_x^{-1} is the differential of f^{-1} at $f(x)$. For if $x, y \in G$, we may write

$$f^{-1}[f(y)] - f^{-1}[f(x)] = l_x^{-1}(f(y) - f(x)) + R'(f(x), f(y)).$$

We need only show that

$$\lim_{|f(y) - f(x)| \to 0} \frac{|R'(f(x), f(y))|}{|f(y) - f(x)|} = 0.$$

But the above expression is simply

$$y - x = l_x^{-1}[l_x(y - x) + R(x, y)] + R'(f(x), f(y)),$$

where

$$\lim_{|y - x| \to 0} \frac{|R(x, y)|}{|y - x|} = 0.$$

We then have

$$R'(f(x), f(y)) = -l_x^{-1}[R(x, y)],$$

and since there is a K such that

$$|l_x^{-1}(z)| \le K|z|,$$

for every $z \in F$, we have

$$|R'(f(x), f(y))| \le K|R(x, y)|.$$

It follows that

$$\frac{|R'(f(x), f(y))|}{|f(y) - f(x)|} \le K \frac{|R(x, y)|}{|y - x|} \cdot \frac{|y - x|}{|f(y) - f(x)|}.$$

But there is a K' such that $|x - y| < K'|l_x(x - y)|$, and for $|y - x|$ sufficiently small,

$$|f(y) - f(x)| > \tfrac{1}{2}|l_x(y - x)|.$$

The result readily follows.

Now, f is continuous, open, and one-one, so that it is a homeomorphism between G and $f(G)$. The mapping

$$Df: G \to L(E, F)$$

is continuous. The mapping ϕ, defined in (c) of Section 1, which takes each regular $l \in L(E, F)$ into its inverse $l^{-1} \in L(F, E)$, is a homeomorphism. Let

$$Df^{-1}: f(G) \to L(E, F)$$

be the differential of f^{-1}. Since

$$(Df^{-1})[f(x)] = [Df(x)]^{-1},$$

we have

$$Df^{-1} = \phi \circ Df \circ f^{-1}.$$

But the continuity of f^{-1}, Df, and ϕ implies the continuity of Df^{-1}. ∎

In summary, we have shown that if f is continuously differentiable on an open set G and is regular at an $x \in G$, there is an open set N, $x \in N$, on which f is a homeomorphism with a continuously differentiable, regular inverse.

3. IMPLICIT FUNCTION THEOREM

The results of the last section may be used to prove a general implicit function theorem. However, it seems to be worthwhile and instructive to first give a special proof for the one-dimensional case.

If we have an expression $f(x, y) = 0$, it may express y implicitly as a function of x. For example, let

$$f(x, y) = x^2 + y^2 - 4 = 0.$$

The point $(0, 2)$ satisfies this relation, i.e., $f(0, 2) = 0$, and there is an open interval I, on the x-axis, to which $x = 0$ belongs, and a function $\phi: I \to R$ such that $\phi(0) = 2$ and $f(x, \phi(x)) = 0$, for all $x \in I$.

However, we also have $f(2, 0) = 0$, but there is no analogous interval and function corresponding to this point.

We have

Theorem 4

Let $G \subset R \times R$ be open and let

$$f: G \to R$$

be continuously differentiable. Let $(x_0, y_0) \in G$ be such that

$$f(x_0, y_0) = 0 \quad \text{and} \quad \frac{\partial f}{\partial y}(x_0, y_0) \neq 0.$$

Then there is an open interval $I \subset R$, such that $x_0 \in I$, and there is a function

$$\phi : I \to R$$

which is continuously differentiable, such that $(x, \phi(x)) \in G$, for all $x \in I$, $\phi(x_0) = y_0$, and such that

$$f(x, \phi(x)) = 0, \qquad \text{for all } x \in I.$$

Proof

We may suppose that

$$\frac{\partial f}{\partial y}(x_0, y_0) > 0.$$

There is a closed rectangle, $R = I' \times J$, with center (x_0, y_0), such that

$$\frac{\partial f}{\partial y}(x, y) > 0,$$

for all $(x, y) \in R$. Let $J = [c, d]$. Then

$$f(x_0, c) < 0 < f(x_0, d).$$

By continuity of f, there is an open $I \subset I'$, with center x_0, such that $f(x, c) < 0$, for every $x \in I$, and $f(x, d) > 0$, for every $x \in I$.

Since f is continuous in y, for every $x \in I$ there is a unique (since f is monotone on $[c, d]$ in y) $\phi(x)$, such that $f(x, \phi(x)) = 0$ and $c < \phi(x) < d$.

We show that ϕ is continuous on I. Let $\xi \in I$ and let $\epsilon > 0$. By the above argument applied to $(\xi, \phi(\xi))$, there is an interval $(\xi - \delta, \xi + \delta)$ such that for every $x \in (\xi - \delta, \xi + \delta)$, we have $|\phi(x) - \phi(\xi)| < \epsilon$.

It remains for us to show that ϕ is continuously differentiable on I. For $(x, y) \in G$, $(x + h, y + k) \in G$, such that the line segment joining them is contained in G, there is a (ξ, η) on this line segment such that

$$f(x + h, y + k) - f(x, y) = \frac{\partial f}{\partial x}(\xi, \eta) \cdot h + \frac{\partial f}{\partial y}(\xi, \eta) \cdot k.$$

Now, let $x \in I$ and $x + h \in I$. Then

$$0 = f(x + h, \phi(x + h)) - f(x, \phi(x))$$

$$= \frac{\partial f}{\partial x}(\xi, \eta) \cdot h + \frac{\partial f}{\partial y}(\eta, \xi)[\phi(x + h) - \phi(x)],$$

where (ξ, η) is on the line segment joining $(x, \phi(x))$ to $(x + h, \phi(x + h))$. Then

$$\frac{\phi(x + h) - \phi(x)}{h} = -\frac{\dfrac{\partial f}{\partial x}(\xi, \eta)}{\dfrac{\partial f}{\partial y}(\xi, \eta)}.$$

Since $\partial f/\partial y \neq 0$ in a neighborhood of $(x, \phi(x))$ and since the two partial derivatives are continuous, it follows that the limit exists, as h tends to zero. Thus $\phi'(x)$ exists and

$$\phi'(x) = -\frac{\dfrac{\partial f}{\partial x}(x, \varphi(x))}{\dfrac{\partial f}{\partial y}(x, \varphi(x))}. \quad \blacksquare$$

This theorem may be extended to mappings from $(m + n)$-dimensional space into n-dimensional space. Using the results of Section 2, we now prove a general implicit function theorem.

Theorem 5

Let E be an m-dimensional, K an n-dimensional, and F an n-dimensional vector space. Let $G \subset E \times K$ be an open set. (For each $x \in E$, $y \in K$, there is a point $(x, y) \in E \times K$.) Let

$$f: G \to F$$

be continuously differentiable and let $(x^0, y^0) \in G$ be such that

(a) $f(x^0, y^0) = \theta$,

(b) *the differential l of f at (x^0, y^0) is such that $l(\theta, y) = \theta$ implies $y = \theta$. (This condition corresponds to $\partial f/\partial y(x^0, y^0) \neq 0$, for the case $n = m = 1$.)*

There is then a neighborhood N of x^0 in E and a mapping,

$$\phi: N \to K,$$

which is continuously differentiable, for which $[(x, \varphi(x)): x \in N] \subset G$, and such that

$$f(x, \phi(x)) = \theta,$$

for all $x \in N$.

Proof

The mapping f induces a mapping

$$\psi: G \to E \times F,$$

defined by $\psi(x, y) = (x, f(x, y))$.

We first note that ψ is continuously differentiable and regular at (x^0, y^0). For this, we write

$$\psi(x, y) - \psi(x^0, y^0) = (x, f(x, y)) - (x^0, f(x^0, y^0))$$
$$= (x, f(x, y)) - (x^0, \theta) = (x - x^0, f(x, y)).$$

It should be clear from this that the differential of ψ at (x^0, y^0) exists and is the element

$$\lambda \in L(E \times K, E \times F),$$

for which

$$\lambda(u, v) = (u, l(u, v)).$$

The continuous differentiability of ϕ and the above relation between the differentials of ϕ and of ψ imply the continuous differentiability of ψ.

Moreover, ψ is regular. For $u \neq \theta$ implies $(u, l(u, v)) \neq (\theta, \theta)$, and by **(b)**, $u = \theta$, $v \neq \theta$ implies $(u, l(u, v)) \neq (\theta, \theta)$.

We may thus apply the results of Section 2 to the mapping ψ. Accordingly, there are open sets $U \subset G$ and $V \subset E \times F$ such that $(x^0, y^0) \in U$ and

$$\psi: U \to V$$

is a homeomorphism, with ψ^{-1} continuously differentiable. Now, on V,

$$\psi^{-1}(x, z) = (x, h(x, z)),$$

where h is continuously differentiable on V into K. Then, for $(x, z) \in V$ and $(x, h(x, z)) \in U$,

$$f(x, h(x, z)) = z,$$

since $\psi(x, h(x, z)) = (x, z)$ and by definition

$$\psi(x, h(x, z)) = (x, f(x, h(x, z))).$$

Let N be a neighborhood of x^0 such that $N \times \{\theta\} \subset V$, and let

$$\phi(x) = h(x, \theta),$$

for all $x \in N$. Then $(x, \phi(x)) \in U$, for all $x \in N$. Since

$$f(x, h(x, \theta)) = \theta,$$

we have

$$f(x, \phi(x)) = \theta,$$

for all $x \in N$.

Moreover, we have shown that ϕ is continuously differentiable and that

$$\phi(x^0) = h(x^0, 0) = y^0,$$

since

$$\psi(x^0, y^0) = (x^0, 0). \quad \blacksquare$$

The proof of Theorem 5 is now complete.

If there are coordinates $(x_1, \cdots, x_m)$ in E, $(u_1, \cdots, u_n)$ in K, and $(y_1, \cdots, y_n)$ in F, the mapping f may be written out as

$$y_1 = f_1(x_1, \cdots, x_m; u_1, \cdots, u_n)$$
$$\cdots \cdots$$
$$y_n = f_n(x_1, \cdots, x_m; u_1, \cdots, u_n).$$

We consider the system

$$f_1(x_1, \cdots, x_m; u_1, \cdots, u_n) = 0$$
$$\cdots \cdots$$
$$f_n(x_1, \cdots, x_m; u_1, \cdots, u_n) = 0,$$

and suppose it is satisfied at $(x^0, u^0) = (x_1^0, \cdots, x_m^0 ; u_1^0, \cdots, u_m^0)$.

We shall find what condition (b) means, in the statement of Theorem 5. The differential of f at (x^0, u^0) is given by the matrix

$$\begin{pmatrix} \dfrac{\partial f_1}{\partial x_1}(x^0, u^0) \cdots \dfrac{\partial f_1}{\partial x_m}(x^0, u^0) & \dfrac{\partial f_1}{\partial u_1}(x^0, u^0) \cdots \dfrac{\partial f_1}{\partial u_n}(x^0, u^0) \\ \cdots \\ \dfrac{\partial f_n}{\partial x_1}(x^0, u^0) \cdots \dfrac{\partial f_n}{\partial x_m}(x^0, u^0) & \dfrac{\partial f_n}{\partial u_1}(x^0, u^0) \cdots \dfrac{\partial f_n}{\partial u_n}(x^0, u^0) \end{pmatrix},$$

and it is the matrix of the linear mapping

$$y_1 = \frac{\partial f_1}{\partial x_1}(x^0, u^0)x_1 + \cdots + \frac{\partial f_1}{\partial x_m}(x^0, u^0)x_m + \frac{\partial f_1}{\partial u_1}(x^0, u^0)u_1$$
$$+ \cdots + \frac{\partial f_1}{\partial u_n}(x^0, u^0)u_n$$
$$\cdots$$
$$y_n = \frac{\partial f_n}{\partial x_1}(x^0, u^0)x_1 + \cdots + \frac{\partial f_n}{\partial x_m}(x^0, u^0)x_m + \frac{\partial f_n}{\partial u_1}(x^0, u^0)u_1$$
$$+ \cdots + \frac{\partial f_n}{\partial u_n}(x^0, u^0)u_n.$$

If $(x_1, \cdots, x_m) = (0, \cdots, 0)$, the mapping becomes

$$y_1 = \frac{\partial f_1}{\partial u_1} (x^0, u^0)u_1 + \cdots + \frac{\partial f_1}{\partial u_n} (x^0, u^0)u_n$$

(*) $\cdots$

$$y_n = \frac{\partial f_n}{\partial u_1} (x^0, u^0)u_1 + \cdots + \frac{\partial f_n}{\partial u_n} (x^0, u^0)u_n.$$

Thus, condition **(b)** merely asserts that the element of $L(K, F)$ defined by the matrix

$$\begin{pmatrix} \dfrac{\partial f_1}{\partial u_1} (x^0, u^0) \cdots \dfrac{\partial f_1}{\partial u_n} (x^0, u^0) \\ \cdots \\ \dfrac{\partial f_n}{\partial u_1} (x^0, u^0) \cdots \dfrac{\partial f_n}{\partial u_n} (x^0, u^0) \end{pmatrix}$$

is regular. In other words, the rows of this, the Jacobian matrix of the transformation defined by (*), are linearly independent.

4. DETERMINANT, ORIENTED VOLUME

The above facts may also be expressed using the notion of determinant of a matrix. Moreover, for facts concerning the change in volume of a set under a transformation, the determinant is indispensable.

Indeed, we shall discuss a notion of oriented volume along with that of determinant. An n-dimensional space has two orientations. In the case of one dimension, each orientation is given by a unit vector. Thus, if the orientation given the space by the unit vector e is taken to be positive, then the one given by the vector $-e$ is negative. In n-space, the situation is more complicated, although there are still two orientations. Thus, let

$$e_1, e_2, \cdots, e_n$$

be an ordered set of n orthonormal vectors in euclidean n-space. Suppose the positive orientation is given to the space by this set of vectors (this is mere convention, but it determines the orientation given to the space by any linearly independent ordered set of n vectors). Let

$$e_{i_1}, \cdots, e_{i_n}$$

be any reordering of our set $e_1, \cdots, e_n$. The orientation which it gives the space is positive if it may be obtained from $e_1, \cdots, e_n$ by an even number of interchanges of adjacent pairs of vectors. If an odd number of such interchanges is required, the orientation which it gives the space is negative.

If for any vector in the set, its negative is substituted, the orientation of the space is changed. We shall discuss later the orientation given to the space by other sets of vectors.

We shall define a real function $d(A) = |A|$ on the set of all $n \times n$ matrices, called the determinant of A.

The **determinant** of a matrix is defined as follows:

(a) If $A = I$, the identity matrix, where $a_{ij} = \delta_{ij}$, then

$$d(I) = 1.$$

(b) If A and B are such that B is obtained from A by interchanging two adjacent rows, then

$$d(B) = -d(A).$$

(c) If A and B are such that B is obtained from A by multiplying one of the rows by a constant c, then

$$d(B) = cd(A).$$

(d) If A and B are such that B is obtained from A by adding one row of A to another row of A, leaving the other $n - 1$ rows unchanged, then

$$d(B) = d(A).$$

Since, as is easy to show and should be known to the reader, every matrix may be obtained, in a finite number of steps, by starting with the identity matrix and applying the elementary operations described by **(b)**, **(c)**, and **(d)**, the above properties uniquely define the determinant for all $n \times n$ matrices (see Exercise 4.2).

The determinant function also has the following properties, whose proofs are also assumed to be known. (The reader who is unfamiliar with these facts can find them in Birkhoff and MacLane.)

(α) $d(A) \neq 0$ if and only if the vectors given by the rows of A (the row vectors of A) are linearly independent.

(β) For every A and B,

$$d(AB) = d(A)\, d(B).$$

In particular, even though in general $AB \neq BA$, we always have $d(AB) = d(BA)$.

(γ) For a matrix

$$A = \begin{pmatrix} a_{11} & a_{12} & \cdots & a_{1n} \\ a_{21} & a_{22} & \cdots & a_{2n} \\ \cdots & & & \\ a_{n1} & a_{n2} & \cdots & a_{nn} \end{pmatrix},$$

we have

$$d(A) = \sum \sigma(i_1, \cdots, i_n) a_{1i_1} a_{2i_2} \cdots a_{ni_n},$$

where the sum is over all permutations $(i_1, \cdots, i_n)$ of $(1, \cdots, n)$ and $\sigma(i_1, \cdots, i_n) = +1$ if the number of interchanges of adjacent pairs needed to transpose $(1, \cdots, n)$ into $(i_1, \cdots, i_n)$ is even, and $\sigma(i_1, \cdots, i_n) = -1$ if this number of interchanges is odd.

The sign of $d(A)$ is defined to be the orientation given the space by the ordered set of vectors forming the rows of A. For orthonormal vectors, this is easily seen to agree with the orientation discussed above. We now show that the magnitude of $d(A)$ is the volume of the parallelopiped determined by the row vectors of A.

Let $a^1, \cdots, a^n$ be vectors in euclidean space R^n. The **parallelopiped** determined by these vectors is the set of all

$$x = t_1 a^1 + \cdots + t_n a^n,$$

where $0 \leq t_i \leq 1$, $i = 1, \cdots, n$.

The parallelopiped determined by $a^1, \cdots, a^n$ has a volume

$$v(a^1, \cdots, a^n).$$

If we consider the $(n - 1)$-dimensional volume $v(a^2, \cdots, a^n)$ of the parallelopiped determined by $n - 1$ of the vectors, then it can be shown that

$$v(a^1, \cdots, a^n) = v(a^2, \cdots, a^n) \cdot k(a^1),$$

where $k(a^1)$ is the length of the component of the vector a^1 orthogonal to the space determined by $a^2, \cdots, a^n$. (If the set $a^2, \cdots, a^n$ is linearly dependent, we take $k(a^1) = 1$, and in this case, $v(a^1, \cdots, a^n) = v(a^2, \cdots, a^n) = 0$.)

It follows that

$$v(a^1 + a^2, a^2, \cdots, a^n) = v(a^1, a^2, \cdots, a^n),$$

since obviously $k(a^1 + a^2) = k(a^1)$.

Since a matrix A has n-row vectors, we have the function $v(A)$ of matrices, where $v(A) = v(a^1, \cdots, a^n)$ for $a^i = (a_{i1}, \cdots, a_{in})$, $i = 1, \cdots, n$. This volume function has the properties:

(a') $v(I) = 1$, where I is the identity matrix.

(b') If A and B are such that B is obtained from A by interchanging two adjacent rows, then
$$v(B) = v(A).$$

(c') If A and B are such that B is obtained from A by multiplying one of the rows by a constant c, then
$$v(B) = |c|\, v(A).$$

(d') If A and B are such that B is obtained from A by adding one row of A to another row of A and by leaving the other $n - 1$ rows of A unchanged, then
$$v(B) = v(A).$$

Now, if we define the **oriented volume** of an oriented parallelopiped determined by an ordered set of n-vectors to have the magnitude of the volume and the sign of the orientation given the space by the ordered set of vectors, it follows that the oriented volume, $\tilde{v}(A)$, satisfies
$$\tilde{v}(A) = d(A),$$
for every matrix A.

Let E and F be n-dimensional vector spaces and let $l \in L(E, F)$. Let $(x_1, \cdots, x_n)$ be orthonormal coordinates in E and $(y_1, \cdots, y_n)$ in F. Then l is given by a matrix A. We show the effect of l on volumes. Consider an oriented parallelopiped in E determined by the ordered set $b_1, \cdots, b_n$ of vectors, and so by the matrix $B = (b_{ij})$, where $b_i = (b_{i1}, \cdots, b_{in})$, $i = 1, \cdots, n$. This parallelopiped is transformed by A into the oriented parallelopiped of matrix AB. Thus, if the oriented parallelopiped of matrix C is designated as $\tilde{C}$, we have
$$\tilde{v}(l(\tilde{B})) = \tilde{v}(\widetilde{AB}) = d(AB) = d(A)\,d(B) = d(A)\tilde{v}(\tilde{B}).$$

Proposition 2

If $S \subset E$ is an oriented open set with orientation induced by that of the space, then
$$\tilde{v}(l(S)) = d(A)\tilde{v}(S),$$
$$v(l(S)) = |d(A)|\, v(S).$$
If $v(S) = 0$, then $v(l(S)) = 0$.

Proof

Consider a partitioning of S into nonoverlapping intervals and apply the above considerations to each of the intervals. ∎

5. CHANGE OF VARIABLES IN INTEGRATION

In this section, we extend the above result to nonlinear mappings. The crucial point in our argument is

Lemma 2

Let $G \subset E$ be open and let

$$f: G \to F$$

be continuously differentiable, one-one, and regular. For $x \in G$ and $h > 0$, let $I_h(x)$ be the cubical interval of center x and side h, and let $J(x)$ be the determinant of the Jacobian matrix of f at x. Then

$$|J(x)| = \lim_{h \to 0} \frac{v(f(I_h(x)))}{v(I_h(x))},$$

and the convergence is uniform in x on every compact $K \subset G$.

Remark 1 In other words, for every $\epsilon > 0$, there is a $\delta > 0$ such that $x \in K$ and $h < \delta$ implies

$$\left| |J(x)| - \frac{v(f(I_h(x)))}{v(I_h(x))} \right| < \epsilon.$$

Remark 2 $J(x)$ is called the **Jacobian** of f at x.

Proof

Let l_x be the differential of f at x; then l_x^{-1} is the differential of f^{-1} at $f(x)$. Since $f(K)$ is compact, l_x^{-1} is uniformly continuous on $f(K)$. There is thus an M such that

$$|l_x^{-1}| < M$$

for each $x \in K$. Now

$$f(y) - f(x) = l_x(y - x) + R(x, y),$$

where

$$\lim_{|y-x| \to 0} \frac{|R(x, y)|}{|y - x|} = 0,$$

uniformly for $x \in K$. Then

$$l_x^{-1} f(y) - l_x^{-1} f(x) = l_x^{-1} l_x(y - x) + l_x^{-1} R(x, y),$$

where

$$\lim_{|y-x| \to 0} \frac{|l_x^{-1} R(x, y)|}{|y - x|} = 0,$$

uniformly for $x \in K$. Thus

$$\frac{|l_x^{-1} f(y) - l_x^{-1} f(x) - (y - x)|}{|y - x|}$$

converges to zero, uniformly for $x \in K$.

For each $h > 0$, there is then an $\epsilon(h) > 0$ such that

$$\lim_{h \to 0} \epsilon(h) = 0,$$

and for each $x \in K$ and each y on the boundary of $I_h(x)$,

$$l_x^{-1} f(y)$$

lies inside the cubical interval

$$U = I_{h(1+\epsilon(h))} l_x^{-1} f(x)$$

and outside the cubical interval

$$V = I_{h(1-\epsilon(h))} l_x^{-1} f(x).$$

Thus $f(I_h(x))$ satisfies

$$V \subset l_x^{-1} f(I_h(x)) \subset U,$$

since otherwise it would have a boundary point either in V or outside of U, contrary to the above statement. It follows that

$$\lim_{h \to 0} \frac{v(l_x^{-1} f(I_h(x)))}{v(I_h(x))} = 1,$$

uniformly in x on K. But, by Proposition 2,

$$v(l_x^{-1} f(I_h(x))) = \frac{1}{|J(x)|} v(f(I_h(x))).$$

Hence

$$\lim_{h \to 0} \frac{v(f(I_h(x)))}{v(I_h(x))} = |J(x)|,$$

and the convergence is uniform on K. ∎

By using this lemma, we may easily prove

Theorem 6

If $G \subset E$ is open and

$$f: G \to F$$

is continuously differentiable, one-one, and regular, then

$$v(f(G)) = \int_G |J|.$$

Proof

Let $I \subset G$ be a cubical interval. Then I is compact. Let

$$\epsilon_n > 0, \qquad n = 1, 2, \cdots$$

and

$$\lim_{n \to \infty} \epsilon_n = 0.$$

By Lemma 2, there is a partitioning of I into cubical intervals $I_1, \cdots, I_k$, with centers $x^1, \cdots, x^k$, such that the norm of the partition is less than $1/n$ and

$$\left| \frac{v(f(I_j))}{v(I_j)} - |J(x^j)| \right| < \epsilon_n, \qquad j = 1, \cdots, k.$$

It follows that

$$v(f(I)) = \sum_{j=1}^{k} v(f(I_j)) = \sum_{j=1}^{k} |J(x^j)|\, v(I_j) + \epsilon_n \cdot \theta_n \cdot v(I),$$

where $-1 \le \theta_n \le 1$, $n = 1, 2, \cdots$.

Now, as n converges to infinity, the sum

$$\sum_{j=1}^{k} |J(x^j)|\, v(I_j)$$

converges to

$$\int_I |J|$$

and $\epsilon_n \theta_n v(I)$ converges to zero, so that the theorem is proved for I.

Now, G is the union of nonoverlapping cubical intervals $J_1, J_2, \cdots$, $J_k, \cdots$. It follows that

$$v(f(G)) = \sum_{k=1}^{\infty} v(f(J_k)) = \sum_{k=1}^{\infty} \int_{J_k} |J| = \int_G |J|. \quad \blacksquare$$

As a consequence of this theorem, we have

Theorem 7

If $G \subset E$ is open, if

$$f : G \to F$$

is continuously differentiable, one-one, and regular, and if $g : f(G) \to R$ is integrable, then

$$\int_{f(G)} g = \int_G J \cdot g \circ f,$$

i.e.,

$$\int_{f(G)} g(y_1, \cdots, y_n)\, dy_1, \cdots, dy_n$$
$$= \int_G g(f_1(x_1, \cdots, x_n), \cdots, f_n(x_1, \cdots, x_n))$$
$$\times J(x_1, \cdots, x_n)\, dx_1 \cdots dx_n,$$

where J is the Jacobian of f.

Proof

We shall prove the theorem for the case where g is nonnegative. The more general case can then easily be obtained by the reader. Moreover, J is either always positive or always negative. We suppose it is positive.

Suppose g is the constant function 1. By Theorem 6,

$$v(f(G)) = \int_G J = \int_G J \cdot g \circ f.$$

But

$$v(f(G)) = \int_{f(G)} 1 = \int_{f(G)} g,$$

so that the theorem holds for this case.

Let $I_1, I_2, \cdots$ be a partitioning of G into cubical intervals and let $K_j = f(I_j)$, $j = 1, 2, \cdots$. Let g be a real function on $f(G)$ which is equal to a nonnegative constant c_j on the interior of K_j, $j = 1, 2, \cdots$. Then, using the fact that the volume of the boundary of each K_j is zero,

$$\int_{f(G)} g = \sum_{j=1}^{\infty} c_j v(K_j) = \sum_{j=1}^{\infty} c_j \int_{I_j} J = \int_G J \cdot g \circ f,$$

so that the theorem holds for this case.

Now let g be integrable, nonnegative and let

$$\int_{f(G)} g$$

be finite. Let $\epsilon > 0$. There is a partitioning of G, as above, into intervals $I_1, I_2, \cdots$, with corresponding $K_1, K_2, \cdots$, and constants $c_j, d_j, j = 1, 2, \cdots$, such that $c_j \le g(y) \le d_j$ for all $y \in K_j$, $j = 1, 2, \cdots$, and

$$\sum_{j=1}^{\infty} d_j v(K_j) < \sum_{j=1}^{\infty} c_j v(K_j) + \epsilon.$$

In order to assure the existence of such a partitioning, we first note that for every closed interval $I \subset G$ and $\epsilon > 0$, there is a $\delta > 0$, such that, for every partitioning $I_1, \cdots, I_k$ of I of norm less than δ and $\xi_j \in I_j, j = 1, \cdots, k$,

$$\left| \int_{f(I)} g - \sum_{j=1}^{k} g(\xi_j) v(f(I_j)) \right| < \epsilon.$$

(The proof of this simple extension of Theorem 12, Chapter III, is left to the reader.)

Now, let $J_1, \cdots, J_m, \cdots$ be an arbitrary partitioning of G. For each $m = 1, 2, \cdots$, there is then a partitioning

$$I_{\nu_{m-1}+1}, \cdots, I_{\nu_m} \text{ of } J_m$$

such that

$$\sum_{j=\nu_{m-1}+1}^{\nu_m} \sup \left[|g(x) - g(y)| : x, y \in f(I_j) \right] v(f(I_j)) < \frac{\epsilon}{2^m}.$$

The partitioning $I_1, I_2, \cdots$ of G obtained in this way has the required property.

It follows that

$$\overline{\int_G} J \cdot g \circ f - \epsilon < \int_{f(G)} g < \int_G J \cdot g \circ f + \epsilon.$$

Since this holds for every $\epsilon > 0$, we have

$$\int_G J \cdot g \circ f = \int_{f(G)} g. \quad \blacksquare$$

6. APPLICATION TO PROBABILITY

As an application of the transformation of integral formula of the last section, we choose to obtain some simple probability density functions. Our discussion will necessarily be incomplete, but in view of the importance of the subject we prefer to do this than to have more trivial, although perhaps more complete, examples.

Distribution functions and probability density functions are associated with random variables. The notion of a random variable is outside the scope of this work. Nevertheless, we can talk about their distribution functions and probability density functions as the rudimentary objects.

Thus, if a random variable X assumes real values, we have as our starting point, for each real x, the probability that $X < x$, which we write as

$$P(X < x) = \Phi(x).$$

The function Φ is called the **distribution function** of the random variable X. It is a monotonically nondecreasing function such that

$$\lim_{x \to -\infty} \Phi(x) = 0$$

and

$$\lim_{x \to \infty} \Phi(x) = 1.$$

The special cases of interest here are those for which Φ is continuous. More than this, we require that there is a function ϕ such that

$$\Phi(x) = \int_{-\infty}^{x} \phi(t)\, dt.$$

The function ϕ is called the **probability density function** of the random variable X. In particular, if ϕ itself is continuous, then

$$\phi(x) = \Phi'(x).$$

All of our random variables will be assumed to have probability density functions.

If X and Y are random variables, they have a **joint distribution function** and, perhaps, a **joint probability density function**. The joint distribution function is the function

$$\Phi(x, y) = P[X < x,\ Y < y].$$

The joint probability density function, if it exists, is the one, $\phi(x, y)$, for which

$$\Phi(x, y) = \int_{-\infty}^{y} \int_{-\infty}^{x} \phi(u, v)\, du\, dv.$$

Random variables X and Y are called **independent** if their joint probability density function has the form

$$\phi(x, y) = \phi_1(x)\phi_2(y),$$

i.e., the joint probability density function is the product of the probability density functions of X and of Y.

The above definitions may be extended, in an obvious way, to the case of n random variables.

The joint probability density function allows us to find the probability density functions of various new random variables.

If X and Y are random variables, we first wish to consider the random variable $X + Y$. We suppose X has probability density function $f(x)$, that Y has probability density function $g(y)$, and that X and Y are independent random variables. We shall determine the probability density function of the random variable $X + Y$.

Let $H(u)$ be the distribution function of the random variable $X + Y$. Then

$$H(u) = P(X + Y < u) = \iint\limits_{x+y \leq u} f(x)g(y)\, dx\, dy$$

$$= \int_{-\infty}^{\infty} \int_{-\infty}^{u-x} f(x)g(y)\, dx\, dy.$$

We effect the transformation ψ given by

$$\psi: \begin{array}{l} x = \xi \\ y = v - \xi. \end{array}$$

The Jacobian of ψ is then

$$\begin{vmatrix} \dfrac{\partial x}{\partial \xi} & \dfrac{\partial x}{\partial v} \\[2mm] \dfrac{\partial y}{\partial \xi} & \dfrac{\partial y}{\partial v} \end{vmatrix} = \begin{vmatrix} 1 & 0 \\ -1 & 1 \end{vmatrix} = 1.$$

By the transformation of integral formula, we have

$$H(u) = \int_{-\infty}^{\infty} \int_{-\infty}^{u-x} f(x)g(y)\, dx\, dy = \int_{-\infty}^{\infty} \int_{-\infty}^{u} f(\xi)g(v - \xi)\, d\xi\, dv$$

$$= \int_{-\infty}^{u} dv \int_{-\infty}^{\infty} f(\xi)g(v - \xi)\, d\xi.$$

We thus see that the probability density function of the random variable $X + Y$ is

$$h(v) = \int_{-\infty}^{\infty} f(x)g(v - x)\, dx.$$

A random variable X is called **positive** if $P(X < 0) = 0$. Let X and Y be independent random variables, with X positive. We want the probability density function of the random variable Y/X.

Now, $P(Y/X < u)$ is given by

$$H(u) = \iint\limits_{y/x < u, x > 0} f(x)g(y)\, dx\, dy = \int_{0}^{\infty} \int_{-\infty}^{ux} f(x)g(y)\, dx\, dy,$$

where $f(x)$ and $g(y)$ are the probability density functions of X and Y, respectively.

We effect the transformation ψ given by

$$\psi: \begin{array}{l} x = \xi \\ y = v\xi. \end{array}$$

The Jacobian of ψ is then

$$\begin{vmatrix} \dfrac{\partial x}{\partial \xi} & \dfrac{\partial x}{\partial v} \\[2mm] \dfrac{\partial y}{\partial \xi} & \dfrac{\partial y}{\partial v} \end{vmatrix} = \begin{vmatrix} 1 & 0 \\ v & \xi \end{vmatrix} = \xi.$$

By the transformation of integral formula,

$$H(u) = \int_0^\infty \int_{-\infty}^{ux} f(x)g(y)\,dx\,dy = \int_0^\infty \int_{-\infty}^u f(\xi)g(v\xi)\xi\,d\xi\,dv$$

$$= \int_{-\infty}^u dv \int_0^\infty f(\xi)g(v\xi)\xi\,d\xi.$$

The probability density function of Y/X is thus

$$h(v) = \int_0^\infty f(x)g(vx)x\,dx.$$

A random variable may have a mean. If its probability density function is $f(x)$, then its **mean,** or expected value, is

$$\mu = \int_{-\infty}^\infty xf(x)\,dx,$$

if this integral exists. If its mean exists, then the random variable may have a **variance,** which is defined as

$$\sigma^2 = \int_{-\infty}^\infty (x - \mu)^2 f(x)\,dx,$$

if this integral exists.

The most important random variables are those which are normally distributed. The probability density function of a random variable which is normally distributed, with mean μ and variance σ^2, is given by

$$f(x) = \frac{1}{\sqrt{2\pi}\sigma} \exp\left[-\frac{(x - \mu)^2}{2\sigma^2}\right].$$

For example, if we have a sequence X_1, X_2, $\cdots$ of random variables, all having the same probability density functions, with $\mu = 0$, $\sigma^2 = 1$, then the probability density functions of

$$Y_n = \frac{1}{n}(X_1 + X_2 + \cdots + X_n)$$

tend to the normal with $\mu = 0$, $\sigma^2 = 1$.

It is not difficult to show that if X and Y are independent random variables which are normally distributed, with means μ_1, μ_2 and variances σ_1^2, σ_2^2, respectively, then $X + Y$ is normally distributed with mean $\mu_1 + \mu_2$ and variance $\sigma_1^2 + \sigma_2^2$. We leave the proof to the reader.

We now derive the probability density functions of three of the most important random variables of statistics.

(a) Let $X_1, \cdots, X_n$ be normally distributed, independent, all with mean 0 and variance 1. We then consider the random variable

$$Y = X_1^2 + \cdots + X_n^2.$$

This is the so-called χ^2 (chi square) random variable.

In order to find the probability density function of Y, we consider the joint probability density function,

$$\left(\frac{1}{\sqrt{2\pi}}\right)^n \exp\left[-\frac{x_1^2 + \cdots + x_n^2}{2}\right],$$

of $X_1, \cdots, X_n$. Then we have

$$P(Y < r) = \int_{|x|^2 < r} \left(\frac{1}{\sqrt{2\pi}}\right)^n \exp\left[-\frac{x_1^2 + \cdots + x_n^2}{2}\right] dx_1 \cdots dx_n$$

$$= \int_0^{\sqrt{r}} \left(\frac{1}{\sqrt{2\pi}}\right)^n A_t \exp\left[-\frac{t^2}{2}\right] dt,$$

where A_t is the area of the surface bounding the n-ball of radius t. The area function is the derivative of the volume function, and so has the form ct^{n-1}. Using the same c to designate all constant multipliers, we obtain

$$P(Y < r) = c \int_0^{\sqrt{r}} t^{n-1} \exp\left[-\frac{t^2}{2}\right] dt,$$

and the distribution function of Y is

$$F(x) = c \int_0^x u^{(n-1)/2} \exp\left[-\frac{u}{2}\right] u^{-1/2} du = c \int_0^x u^{n/2-1} \exp\left[-\frac{u}{2}\right] du,$$

so that the probability density function of Y is given by

$$f(x) = \begin{cases} cx^{n/2-1} \exp\left[-\dfrac{x}{2}\right] & \text{if } x > 0, \\ 0, & \text{if } x \leq 0. \end{cases}$$

Since

$$\int_0^\infty f(x)\, dx = 1,$$

we find

$$c = \frac{1}{2^{n/2}\Gamma(n/2)}.$$

If the variances of the $X_1, \cdots, X_n$ are σ^2, instead of 1, the probability density function becomes

$$f(x) = \begin{cases} \dfrac{1}{2^{n/2}\sigma^n\Gamma(n/2)}\, x^{n/2-1} \exp\left[-\dfrac{x}{2\sigma^2}\right], & \text{if } x > 0, \\ 0, & \text{if } x \le 0. \end{cases}$$

The related random variable

$$\left[\frac{1}{n}(X_1^2 + \cdots + X_n^2)\right]^{1/2}$$

has the probability density function

$$f(x) = \begin{cases} \dfrac{2(n/2)^{n/2}}{\sigma^n\Gamma(n/2)}\, x^{n-1} \exp\left[-\dfrac{nx^2}{2\sigma^2}\right], & \text{if } x > 0, \\ 0, & \text{if } x \le 0. \end{cases}$$

The proofs are left to the reader.

(b) We next obtain the probability density function for the celebrated Student's t-distribution. For this, let $X, X_1, \cdots, X_n$ be normal, independent, and with mean 0 and variance σ^2. We are interested in the probability density function of the quotient

$$t = \frac{X}{\dfrac{1}{n}\sum_{i=1}^{n} X_i^2}.$$

The numerator of this quotient has the probability density function

$$g(y) = \frac{1}{\sqrt{2\pi}\,\sigma} \exp\left[-\frac{y^2}{2\sigma^2}\right],$$

and the denominator has the probability density function

$$f(x) = \frac{2\left(\dfrac{n}{2}\right)^{n/2}}{\sigma^n\Gamma\left(\dfrac{n}{2}\right)}\, x^{n-1} \exp\left[-\frac{nx^2}{2\sigma^2}\right].$$

The quotient then has the probability density function

$$h(u) = c \int_0^\infty x^{n-1} \exp\left[-\frac{nx^2}{2\sigma^2}\right] \exp\left[-\frac{u^2 x^2}{2\sigma^2}\right] x \, dx$$

$$= c \int_0^\infty x^n \exp\left[-\frac{x^2}{2\sigma^2}(n + u^2)\right] dx.$$

For fixed u, let

$$y = \sqrt{n + u^2} \, x.$$

Then

$$h(u) = c \int_0^\infty y^n \exp\left[-\frac{y^2}{2\sigma^2}\right] \cdot \frac{1}{(n + u^2)^{(n+1)/2}} \, dy$$

$$= c(n + u^2)^{-(n+1)/2}$$

$$= \frac{\Gamma\left(\frac{n+1}{2}\right)}{\Gamma\left(\frac{n}{2}\sqrt{n\pi}\right)} \left(1 + \frac{u^2}{n}\right)^{-(n+1)/2}.$$

Observe that the probability density function of t is independent of the variance. It may be used to test whether the means of two normally distributed random variables are the same, under the assumption that the variances are the same, without assuming what the value of the variance actually is.

(c)   As a final example, we consider the Snedecor F-test, which is of great use in deciding whether variations within batches of a product are as great as are variations between batches.

Let $X = \dfrac{1}{m}(X_1^2 + \cdots + X_m^2)$ and $Y = \dfrac{1}{n}(Y_1^2 + \cdots + Y_n^2)$, where $X_1, \cdots, X_m, Y_1, \cdots, Y_n$ are independent, normal, with mean 0 and variance 1. The F-random variable is then the quotient

$$F = \frac{Y}{X}.$$

Since the probability density functions of Y and X are

$$g(y) = cy^{n/2-1} \exp\left[-\frac{ny}{2}\right] \qquad y > 0,$$

and

$$f(x) = cx^{n/2-1} \exp\left[-\frac{nx}{2}\right], \qquad x > 0,$$

the probability density function of F is

$$
h(u) = c \int_0^\infty x^{m/2-1} \exp\left[-\frac{mx}{2}\right] (xu)^{n/2-1} \exp\left[-\frac{nxu}{2}\right] x\,dx
$$

$$
= cu^{n/2-1} \int_0^\infty x^{(m+n)/2-1} \exp\left[-\frac{x}{2}(m+nu)\right] dx
$$

$$
= cu^{n/2-1}(m+nu)^{-(m+n)/2}
$$

$$
= \frac{\Gamma\left(\dfrac{m+n}{2}\right)}{\Gamma\left(\dfrac{m}{2}\right)\Gamma\left(\dfrac{n}{2}\right)} \left(\frac{n}{m}\right)^{n/2} u^{n/2-1}\left(1+\frac{n}{m}u\right)^{-(m+n)/2}
$$

7. LENGTH AND AREA

In this section, we discuss the length of a curve and the area of a surface. We shall consider curves in euclidean 3-space E. First, let $I = [a, b]$ be a closed interval on the real line and let

$$
f: I \to E
$$

be a continuous mapping. Let $\mathscr{C}$ be the set of all such continuous mappings. We define an equivalence relation in $\mathscr{C}$ by letting f be equivalent to g, $f \sim g$, if there is a mapping

$$
h: I \to I,
$$

which is a homeomorphism, such that

$$
f = g \circ h.
$$

This is an equivalence relation since, as is easy to verify, (1) $f \sim f$, (2) $f \sim g$ implies $g \sim f$, and (3) $f \sim g$, $g \sim k$ implies $f \sim k$.

By a **curve** C, we mean an equivalence class of mappings in $\mathscr{C}$. Each $f \in C$ is called a **representation** of the curve C. In defining the length of a curve C, we consider any $f \in C$. For any partition $\pi = \{a = x_0 < \cdots < x_n = b\}$ of I, consider the number

$$
\lambda(f, \pi) = \sum_{i=1}^n |f(x_i) - f(x_{i-1})|
$$

and let

$$
l(f) = \sup \lambda(f, \pi),
$$

for all partitions π.

This number $l(f)$ is the **length of the curve** C and is also written $l(C)$. In order to justify this definition, it is only necessary to note that $f \sim g$

implies $l(f) = l(g)$. But this follows from the fact that, for every partition π, there is a partition π' such that

$$\lambda(f, \pi) = \lambda(g, \pi').$$

We now give a criterion that a curve should have finite length. A real function u on I is said to be of **bounded variation** if there is an M such that, for every partition $\pi = \{a = x_0 < x_1 < \cdots < x_n = b\}$ of I, we have

$$v(u, \pi) = \sum_{i=1}^{n} |u(x_i) - u(x_{i-1})| < M.$$

We also define the variation of u as the number

$$V(u) = \sup v(u, \pi),$$

for all partitions π. Let $f \in C$ be given by

$$f: \begin{aligned} u_1 &= u_1(x), \\ u_2 &= u_2(x), \\ u_3 &= u_3(x), \end{aligned}$$

in terms of a coordinate system in E. Then, for every partition π of I, it is clear that

$$v(u_i, \pi) \leq \lambda(f, \pi) \leq v(u_1, \pi) + v(u_2, \pi) + v(u_3, \pi), \qquad i = 1, 2, 3,$$

so that

Proposition 3

$l(f)$ *is finite if and only if the coordinate functions* u_1, u_2, u_3 *are all of bounded variation.*

Let $\{\pi_m\}$ be a sequence of partitions of I whose norms converge to zero. We show that

$$\lim_{m \to \infty} \lambda(f, \pi_m)$$

either exists or is $+\infty$ and that this limit is $l(f)$. For this purpose, we define a sequence $\{s_n\}$ of real numbers to be **quasi-monotonically non-decreasing** if, for every $\epsilon > 0$ and n, there is an N such that $m > N$ implies $s_m > s_n - \epsilon$. We then have

Lemma 3

If $\{s_n\}$ *is a quasi-monotonically nondecreasing sequence, it converges to a real number or to* $+\infty$.

Proof

It is evident that $\{s_n\}$ has a lower bound.

Suppose $\{s_n\}$ is unbounded. For every M, there is an n such that $s_n > M + 1$. There is then an N such that $m > N$ implies $s_m > M$. Thus $\{s_n\}$ converges to $+\infty$.

Suppose $\{s_n\}$ is bounded. Let $u = \sup s_n$ and let $\epsilon > 0$. There is an n such that $s_n > u - \epsilon/2$ and an N such that $m > N$ implies $s_m > s_n - \epsilon/2$. Thus $m > N$ implies $u - \epsilon < s_m \leq u$, and $\{s_n\}$ converges to u. ∎

Let $f \colon I \to E$ be continuous and let $\{\pi_m\}$ be a sequence of partitions of I whose norms converge to zero. Fix $\epsilon > 0$ and let

$$\pi_n = \{t_0 = a < t_1 < \cdots < t_k = b\}.$$

Let $\delta > 0$ be such that

$$\delta < \tfrac{1}{3} \min [t_i - t_{i-1} \colon i = 1, \cdots, k],$$

and if $x, y \in I$, with $|x - y| < \delta$, we have

$$|f(x) - f(y)| < \frac{\epsilon}{2k}.$$

Now, let N be such that $m > N$ implies $\|\pi_m\| < \delta$. For every $m > N$, each interval

$$[t_{i-1}, t_i], \ i = 1, \cdots, k,$$

contains points s_i, u_i which are in π_m and are such that

$$s_i < u_i, \qquad |f(u_i) - f(t_i)| < \frac{\epsilon}{2k}$$

and

$$|f(s_i) - f(t_{i-1})| < \frac{\epsilon}{2k}.$$

It follows that

$$\lambda(f, \pi_m) > \lambda(f, \pi_n) - \epsilon,$$

so that $\{\lambda(f, \pi_m)\}$ is quasi-monotonically nondecreasing; hence, it converges to a real number or to $+\infty$.

Let $\{\sigma_m\}$ be another sequence of partitions whose norms converge to zero. Then the norms of the sequence

$$\sigma_1, \pi_1, \sigma_2, \pi_2, \cdots, \sigma_m, \pi_m, \cdots,$$

converge to zero so that

$$\lambda(f, \sigma_1), \lambda(f, \pi_1), \cdots, \lambda(f, \sigma_m), \lambda(f, \pi_m), \cdots,$$

converges to a real number or to $+\infty$. Hence,

$$\lim_{m \to \infty} \lambda(f, \sigma_m) = \lim_{m \to \infty} \lambda(f, \pi_m).$$

Now, if a sequence is quasi-monotonically nondecreasing, then for every

$$n = 1, 2, \cdots, \quad \lim_{m \to \infty} s_m \geq s_n.$$

Let π be any partition of I and let $\sigma_1 = \pi$ in the above sequence $\{\sigma_m\}$. Then

$$\lim_{m \to \infty} \lambda(f, \pi_m) = \lim_{m \to \infty} \lambda(f, \sigma_m) \geq \lambda(f, \pi),$$

so that

$$\lim_{m \to \infty} \lambda(f, \pi_m) \geq l(f).$$

But it is obvious that

$$\lim_{m \to \infty} \lambda(f, \pi_m) \leq l(f).$$

This proves

Theorem 8

For any $f \in C$ and any sequence $\{\pi_m\}$ of partitions whose norms converge to zero,

$$l(f) = \lim_{m \to \infty} \lambda(f, \pi_m).$$

Finally, we show that if f is continuously differentiable, then $l(f)$ is given by the formula

$$l(f) = \int_a^b [\{u'(x)\}^2 + \{v'(x)\}^2 + \{w'(x)\}^2]^{1/2} \, dx,$$

where

$$u = u(x),$$
$$f\colon v = v(x),$$
$$w = w(x).$$

Let $\epsilon > 0$. There is a $\delta > 0$ such that $x, y \in [a, b]$ and $|x - y| < \delta$ implies

(*) $|u'(x) - u'(y)| < \epsilon, \quad |v'(x) - v'(y)| < \epsilon, \quad |w'(x) - w'(y)| < \epsilon.$

Let

$$\pi = \{a = x_0 < x_1 < \cdots < x_n = b\}$$

be a partition of I, of norm less than δ, such that

$$\lambda(f, \pi) > l(f) - \epsilon.$$

For each $i = 1, \cdots, n$ there are $x_{i,1}, x_{i,2}, x_{i,3}$ in the interval (x_{i-1}, x_i), such that

$$u'(x_{i,1}) \cdot (x_i - x_{i-1}) = u(x_i) - u(x_{i-1}),$$
$$v'(x_{i,2}) \cdot (x_i - x_{i-1}) = v(x_i) - v(x_{i-1}),$$
$$w'(x_{i,3}) \cdot (x_i - x_{i-1}) = w(x_i) - w(x_{i-1}).$$

It follows that

$$\lambda(f, \pi) = \sum_{i=1}^{n} [\{u'(x_{i,1})\}^2 + \{v'(x_{i,2})\}^2 + \{w'(x_{i,3})\}^2]^{1/2}(x_i - x_{i-1}).$$

By (*) and the easily proved inequality

$$|(a_1^2 + a_2^2 + a_3^2)^{1/2} - (b_1^2 + b_2^2 + b_3^2)^{1/2}| \le \sum_{i=1} |a_i - b_i|,$$

for $a_1, a_2, a_3, b_1, b_2, b_3$ nonnegative, we obtain

$$\left| [\{u'(x_{i,1})\}^2 + \{v'(x_{i,2})\}^2 + \{w'(x_{i,3})\}^2]^{1/2}(x_i - x_{i-1}) \right.$$
$$\left. - \int_{x_{i-1}}^{x_i} [\{u'(x)\}^2 + \{v'(x)\}^2 + \{w'(x)\}^2]^{1/2} \, dx \right| < 3\epsilon |x_i - x_{i-1}|.$$

We then have

$$\left| \lambda(f, \pi) - \int_a^b [\{u'(x)\}^2 + \{v'(x)\}^2 + \{w'(x)\}^2]^{1/2} \, dx \right| < 3\epsilon(b - a)$$

and

$$\left| l(f) - \int_a^b [\{u'(x)\}^2 + \{v'(x)\}^2 + \{w'(x)\}^2] \, dx \right| < [3(b - a) + 1]\epsilon.$$

Since $\epsilon > 0$ is arbitrary, the following theorem is proved.

Theorem 9

If $f: I \to E$ is continuously differentiable, then

$$l(f) = \int_a^b [\{u'(x)\}^2 + \{v'(x)\}^2 + \{w'(x)\}^2]^{1/2} \, dx,$$

where u, v, w are the coordinate mappings of f.

The area of a surface is much more difficult to handle. Indeed, the areas of inscribed polyhedra do not always converge to the area of the surface in which they are inscribed, so that even the initial definition must be changed.

We start with an example. Consider a right circular cylinder S of radius 1 and height 1. We inscribe a polyhedron in S as follows: Place n equally spaced circles on the cylinder S, at distance $1/n$ apart, and divide each of these circles into m equal arcs, so that the end points of arcs, on each circle, fall directly above, or below, the midpoint of one of the prescribed arcs on an adjacent circle.

Consider the polyhedron formed by triangles, two of whose vertices are end points of a prescribed arc on one of the circles, and the third is the end point of the arc which falls directly above, or below, the mid-

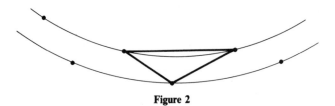

Figure 2

point of this arc on an adjacent circle. (See Fig. 2 for a view of one of the triangles.) These triangles form a polyhedron p_{mn}, inscribed in S, formed by $2mn$ triangles, each of area

$$\sin \frac{\pi}{m} \left[\left(\frac{1}{n}\right)^2 + \left(1 - \cos\frac{\pi}{m}\right)^2 \right]^{1/2}.$$

The area of p_{mn} is then

$$A(p_{mn}) = 2mn \sin \frac{\pi}{m} \left[\left(\frac{1}{n}\right)^2 + \left(1 - \cos\frac{\pi}{m}\right)^2 \right]^{1/2}$$

$$= 2mn \sin \frac{\pi}{m} \left[\left(\frac{1}{n^2} + 4 \sin^4 \frac{\pi}{2m}\right) \right]^{1/2}.$$

Now, as m and n tend to infinity, the polyhedra p_{mn} "tend to S."

We first note that

$$\liminf_{m,n\to\infty} A(p_{mn}) \geq 2\pi.$$

For this,

$$2mn \sin \frac{\pi}{m} \left[\frac{1}{n^2} + 4 \sin^4 \frac{\pi}{2m} \right]^{1/2} \geq 2mn \sin \frac{\pi}{m} \cdot \left(\frac{1}{n}\right) = 2m \sin \frac{\pi}{m},$$

and

$$\lim_{m\to\infty} 2m \sin \frac{\pi}{m} = 2\pi.$$

Next we show that, indeed,
$$\liminf_{m \; n \to \infty} A(p_{mn}) = 2\pi.$$
For this, let $m = n$. Then
$$A(p_{nn}) = 2n^2 \sin \frac{\pi}{n} \left[\frac{1}{n^2} + 4 \sin^4 \frac{\pi}{2n} \right]^{1/2},$$
and
$$\lim A(p_{nn}) = 2\pi.$$
By choosing n large enough compared to m, we can get any limit greater than 2π. The choice $n = m^3$ gives the limit $+\infty$. Indeed,
$$A(p_{mm^3}) = 2m^4 \sin \frac{\pi}{m} \left[\frac{1}{m^6} + 4 \sin^4 \frac{\pi}{2m} \right]^{1/2},$$
and
$$\lim_{m \to \infty} A(p_{mm^3}) = +\infty.$$
The above example shows that a definition analogous to the one for arc length does not work for surface area; it also suggests an appropriate definition.

Let Q be the closed unit square $[0, 1] \times [0, 1]$. We may define the area of a surface given by a continuous mapping
$$f: Q \to R^3,$$
where R^3 is euclidean 3-space.

A triangulation of Q is a subdivision of Q into nonoverlapping triangles such that the vertex of any of the triangles is a vertex of all

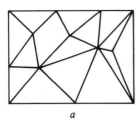

a   b

Figure 3

the others among the triangles to which it may belong. In Fig. 3, (a) is a triangulation while (b) is not.

A mapping $f: Q \to R^3$ is called quasi-linear if it is continuous and if there is a triangulation of Q, on each triangle of which f is affine (i.e., a linear mapping plus a constant).

For a quasi-linear mapping, with triangulation $\Delta = \{\Delta_1, \cdots, \Delta_n\}$, $\Delta_1, \cdots, \Delta_n$ being the triangles of the triangulation, the images $f(\Delta_1), \cdots, f(\Delta_n)$ are triangles in R^3, perhaps degenerate, and so have areas which we write as

$$E(f(\Delta_1)), \cdots, E(f(\Delta_n)).$$

The area of f is defined to be

$$E(f) = \sum_{i=1}^{n} E(f(\Delta_i)).$$

If f is continuous, a quasi-linear p with triangulation $\Delta = \{\Delta_1, \cdots, \Delta_n\}$ is said to be inscribed in f if, for each vertex ξ of any Δ_i, $f(\xi) = p(\xi)$.

We define the area of the surface given by f, by considering for each sequence of quasi-linear $\{p_n\}$, each inscribed in f, the number

$$\liminf_{n} E(p_n),$$

and letting

$$A(f) = \inf \{\liminf_{n} E(p_n)\}$$

be the infimum of these numbers for all such sequences $\{p_n\}$.

A surface is not a mapping, but is an equivalence class of mappings. Here, continuous mappings f and g are equivalent if there is a homeomorphism

$$h: Q \to Q$$

such that

$$f = g \circ h.$$

We leave it for the reader to show that if f is equivalent to g, then

$$A(f) = A(g).$$

We discuss the area $A(f)$ for the case where f is continuously differentiable. Let the coordinates in Q be (u, v) and let the coordinates in R^3 be (x, y, z). Let us consider the functional

$$\Phi(f) = \iint\limits_{Q} [J_1^2(u, v) + J_2^2(u, v) + J_3^2(u, v)]^{1/2} \, du \, dv,$$

where f is given by the mapping

$$x = x(u, v),$$
$$y = y(u, v),$$
$$z = z(u, v),$$

and J_1, J_2, and J_3 are the Jacobians of the mappings $x = x(u, v)$, $y = y(u, v)$ into the (x, y)-plane; $y = y(u, v)$, $z = z(u, v)$ into the (y, z)-plane; and $x = x(u, v)$, $z = z(u, v)$ into the (x, z)-plane, respectively.

It is easy to see that Φ is also defined for quasi-linear mappings p. For this case, the reader may verify that

$$E(p) = \Phi(p).$$

It is not as easy to see, but is also true, that if f is continuously differentiable, if for every n there is a quasi-linear p_n inscribed in f such that $\{p_n\}$ converges uniformly to f, then

$$\liminf_n \Phi(p_n) \geq \Phi(f).$$

Moreover, it is also true, but outside the scope of this work, that there exists a sequence of quasi-linear p_n, each inscribed in f, which converges uniformly to f, such that

$$\lim_n \Phi(p_n) = \Phi(f).$$

Since $\Phi(p_n) = E(p_n)$, $n = 1, 2, \cdots$, it follows from these facts that $\Phi(f) = A(f)$.

The proof for the nonparametric case is easier. In this case, we merely have a continuous real function

$$f: Q \to R,$$

and the area $A(f)$ is defined, just as for mappings, except that quasi-linear approximating functions are used instead of quasi-linear mappings. The function

$$z = f(u, v)$$

may be interpreted as a mapping

$$x = u,$$

$$y = v,$$

$$z = f(u, v).$$

The formula

$$\iint_Q [J_1^2(u, v) + J_2^2(u, v) + J_3^2(u, v)]^{1/2} \, du \, dv$$

reduces, in this case, to

$$\iint\limits_{Q} \left[\left(\frac{\partial f}{\partial u} \right)^2 + \left(\frac{\partial f}{\partial v} \right)^2 + 1 \right]^{1/2} du\, dv.$$

Our method, described above, may be carried out with much less difficulty to show that

$$A(f) = \iint\limits_{Q} \left[\left(\frac{\partial f}{\partial u} \right)^2 + \left(\frac{\partial f}{\partial v} \right)^2 + 1 \right]^{1/2} du\, dv,$$

provided f is continuously differentiable. However, we shall not go on with the details, even though they are within the scope of our experience.

EXERCISES

1.1 Show that the quotient space E/H is indeed a vector space.

1.2 If E is a vector space, H a subspace of E, and $x_1, \cdots, x_k, x_{k+1}, \cdots, x_n$ is a basis in E with $x_1, \cdots, x_k$ a basis in H, show that

$$H_{x_{k+1}}, \cdots, H_{x_n}$$

is a basis in E/H.

2.1 Finish the proof of Lemma 1.

2.2 Show that

$$u = e^x \cos y,$$
$$v = e^x \sin y,$$

is an example of a continuously differentiable mapping which is regular but not one-one.

3.1 Extend the statement and proof of Theorem 4 to the case where

$$G \subseteq R^n \times R$$

and we wish to solve for y in terms of $x_1, \cdots, x_n$.

3.2 In Theorem 4, show that if f is assumed to be of class C^k, then we may conclude that the solution ϕ is of class C^k.

3.3 In this event, find formulas for the second and third derivatives of y with respect to x.

3.4 In the case of Theorem 5, where f is given in coordinates by

$$f_1(x_1, x_2, y_1, y_2), \quad f_2(x_1, x_2, y_1, y_2)$$

find expressions for the partial derivatives of y_1 and y_2 with respect to x_1 and x_2.

3.5 Give conditions under which we may solve the system $f(x, y, z) = 0$, $g(x, y, z) = 0$, for y and z in terms of x.

3.6 If

$$f(x, y, u, v) = 0, \quad g(x, y, u, v) = 0,$$

show that

$$\frac{\partial u}{\partial x} \frac{\partial y}{\partial u} + \frac{\partial v}{\partial x} \frac{\partial y}{\partial v} = 0.$$

3.7 If

$$x_1 = r \cos \phi_1$$
$$x_2 = r \sin \phi_1 \cos \phi_2$$
$$x_3 = r \sin \phi_1 \sin \phi_2 \cos \phi_3$$
$$\cdots$$
$$x_{n-1} = r \sin \phi_1 \cdots \sin \phi_{n-2} \cos \phi_{n-1}$$
$$x_n = r \sin \phi_1 \cdots \sin \phi_{n-1}$$

write the Jacobian

$$\frac{\partial(x_1, \cdots, x_n)}{\partial(r, \phi_1, \cdots, \phi_{n-1})}.$$

3.8 If $f: R^2 \to R^3$ is a differentiable mapping, write the equation of the tangent plane to a point on the surface determined by f.

3.9 Write the equation of the normal to the surface.

4.1 Show that every $n \times n$ matrix can be obtained from the identity matrix by means of the operations described in the text.

4.2 Show that if a matrix A is obtained from the identity in two different ways, by means of these operations, the same value of $d(A)$ is obtained, so that the definition of $d(A)$ is unique.

4.3 Show that $d(A) \neq 0$ if and only if the row vectors of A are linearly independent.

4.4 Show that, for any matrices A and B,

$$d(AB) = d(A) d(B).$$

4.5 Verify the equality

$$v(a^1, \cdots, a^n) = v(a^2, \cdots, a^n) \cdot k(a^1),$$

discussed in the text.

4.6 Give the details of the proof of Proposition 2.

4.7 Let G be open $f: G \to F$ continuously differentiable and regular on G, and $I \subset G$ a closed interval. Show that for every $\epsilon > 0$ there is a $\delta > 0$ such that for every partitioning $I_1, \cdots, I_k$ of I, of norm less than δ, and $\xi_j \in I_j, j = 1, \cdots, k$,

$$\left| \int_{f(I)} g - \sum_{j=1}^{k} g(\xi_j) v(f(I_j)) \right| < \epsilon,$$

where g is integrable on $f(G)$.

5.1 Evaluate

$$\int_{-\infty}^{\infty} e^{-x^2} \, dx$$

by considering

$$\int_{-\infty}^{\infty} e^{-x^2} \, dx \int_{-\infty}^{\infty} e^{-y^2} \, dy$$

as a double integral and making an appropriate change of coordinates.

6.1 If two random variables are independent and have the same probability density function

$$f(x) = \begin{cases} 1, & \text{when} \quad x \in [0, 1], \\ 0, & \text{when} \quad x \notin [0, 1], \end{cases}$$

find the probability density function of their sum.

6.2 Do Exercise 6.1 for the case of 3 independent random variables and for 4 independent random variables.

6.3 If X and Y are independent, positive, and have probability density functions $f(x)$ and $g(y)$, respectively, find the probability density function of XY.

6.4 If X and Y are independent, normal, and with means μ_1, μ_2 and variances σ_1^2, σ_2^2, respectively, show that $X + Y$ is normal with mean $\mu_1 + \mu_2$ and variance $\sigma_1^2 + \sigma_2^2$.

6.5 Derive the formula for

$$\left[\frac{1}{n} (X_1^2 + \cdots + X_n^2) \right]^{1/2}$$

given in the text.

7.1 If $\{f_n\}$ is a sequence of continuous mappings of the unit interval into R^3, which converges uniformly to a mapping f, show that

$$\liminf_{n} l(f_n) \geq l(f).$$

7.2 In the example in the text, given $M > 2\pi$, show that there is a $\phi(m)$ such that

$$\lim_{m \to \infty} A(p_{m\phi(m)}) = M.$$

7.3 Prove that if

$$f: Q \to R$$

is continuously differentiable, then

$$A(f) = \iint_Q \left[\left(\frac{\partial f}{\partial u} \right)^2 + \left(\frac{\partial f}{\partial v} \right)^2 + 1 \right]^{1/2} \, du \, dv.$$

CHAPTER V

MANIFOLDS—DIFFERENTIAL FORMS

Differentiable manifolds are defined and their properties are developed. The tangent space and the space of differentials at a point on a manifold are defined. The algebra of differential forms on a manifold and the operation of exterior differentiation are introduced, leading up to the theorem of Stokes.

1. TOPOLOGICAL SPACE

A **topological space** is a set X, together with a system of subsets $\mathscr{G} = [G]$ of X, called **open** sets, such that

(a)   the set X itself, and the empty set, belong to $\mathscr{G}$;

(b)   the union of sets, belonging to $\mathscr{G}$, belongs to $\mathscr{G}$; i.e., if $G_\alpha \in \mathscr{G}$, $\alpha \in \mathscr{A}$, then

$$\bigcup[G_\alpha : \alpha \in \mathscr{A}] \in \mathscr{G};$$

(c)   if $G_1 \in \mathscr{G}$, $G_2 \in \mathscr{G}$, then $G_1 \cap G_2 \in \mathscr{G}$.

It follows from (c) that the intersection of a finite number of sets in $\mathscr{G}$ is in $\mathscr{G}$.

As examples of topological spaces, we have the following.

Example A   Euclidean n-space, where the open sets are as defined in Chapter I.

Example B The real line, where a set G is open if, for every $x \in G$, x is the left end point of an interval $I \subset G$.

Example C Any set x, with only two open sets, X itself and the empty set.

Example D Any set X, with all subsets of X open. This topology is called the **discrete** topology in X.

Given a topological space X, with open sets $\mathscr{G}$, and given a subset $Y \subset X$, then the set Y, with the sets $\mathscr{H} = [H : H = G \cap Y, G \in \mathscr{G}]$ as open sets, is easily seen to be a topological space. This topology in Y is called the **topology induced by the topology in** X.

An important notion is that of a base for a topology. Let X be a topological space, with open sets $\mathscr{G}$. A subset $\mathscr{B} \subset \mathscr{G}$ is called a **base** for the topology if every $G \in \mathscr{G}$ is the union of sets in $\mathscr{B}$.

For example, for euclidean n-space, the open spheres constitute a base for the topology. For the discrete topology, the empty set, together with all the one-point sets, form a base.

If X, with open sets $\mathscr{G}$, is a topological space, it is easy to see that $\mathscr{B}$ is a base for the topology if and only if for every $G \in \mathscr{G}$ and $x \in G$, there is a $B \in \mathscr{B}$ such that $x \in B$ and $B \subset G$.

A topological space X, with open sets $\mathscr{G}$, is said to be a **Hausdorff space** if, for every $x, y \in X$, $x \neq y$, there are $G, H \in \mathscr{G}$, with $x \in G$ $y \in H$, and $G \cap H$ empty.

All the examples above, with the exception of Example A, are Hausdorff spaces.

A mapping
$$f : X \to Y$$

of a topological space X into a topological space Y is defined to be **continuous** if, for every open set $H \subset Y$, the set $f^{-1}(H)$, of points in X for which $f(x) \in H$, is open in X. The reader should note that the definition given previously for continuity of mappings in euclidean spaces is a special case of this definition.

Finally, a mapping
$$f : X \to Y$$

is called a **homeomorphism** if it is bijective and if f and f^{-1} are both continuous. Topological spaces X and Y are said to be **homeomorphic** if there is a mapping
$$f : X \to Y$$
which is a homeomorphism.

2. MANIFOLD

We are interested in a special class of topological spaces, the manifolds.

An *n*-**dimensional manifold** M is a Hausdorff topological space, such that there is a collection $\mathscr{U} = [U]$ of open sets in M, which covers M, such that each $U \in \mathscr{U}$ is homeomorphic to the open euclidean *n*-ball

$$B_n = [x: x \in R^n, |x| < 1].$$

The statement that $\mathscr{U}$ covers M means that, for every $x \in M$, there is a $U \in \mathscr{U}$ such that $x \in U$.

It is also required by some writers that the manifold be connected. Although most of our examples of manifolds will be connected, it will be necessary, especially for the important Stokes' theorem, to ignore this requirement.

We also note that every *n*-dimensional manifold has a basis whose members are homeomorphic to B_n. We leave the proof to the reader.

We give some examples of manifolds.

Example A   An open interval is a one-dimensional manifold. The collection $\mathscr{U}$ can be taken to have one element, the interval itself, in this case.

Example B   A circle is a one-dimensional manifold. The collection $\mathscr{U}$ can be taken to have two elements (but not one) in this case.

Example C   A closed interval is not a manifold. The end points are not contained in open sets homeomorphic to B_1.

Example D   The space indicated in Fig. 4 is not a manifold. The intersection point p is not contained in any open set homeomorphic to B_1.

Figure 4

Example E   For every n, the *n*-sphere

$$S = [x: x \in R^{n+1}, |x| = 1]$$

is an *n*-dimensional manifold. The collection $\mathscr{U}$ can again be taken to have two elements.

Example F   The torus S is an example of a two-dimensional manifold. It is convenient to represent the torus as a closed square, where points on the top edge and points directly below on the bottom edge are identified (as being the same point). Points on the left and right edges, which are at the same height, are similarly identified (see Fig. 5). In particular, the four vertices are identified as being the same point.

We leave it for the reader to see that $\mathscr{U}$ may be taken to have three elements.

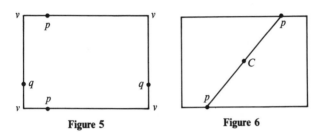

Figure 5   Figure 6

Example G   Another two-dimensional manifold, associated with the square, is the projective plane. In this case, the identification is made for diametrically opposite points. Let c be the center of the square. Every line passing through c cuts the boundary of the square in two points; these are identified (see Fig. 6).

3. DIFFERENTIABLE MANIFOLD

We are not primarily interested in manifolds in full generality. Rather, manifolds which are endowed with a further structure and are called differentiable manifolds, are of great importance to us.

While a manifold is merely a topological space, for a differentiable manifold, the covering family

$$\mathscr{U} = [U_\alpha : \alpha \in \mathscr{A}]$$

and the special mappings

$$\phi_\alpha : U_\alpha \to B_n, \qquad \alpha \in \mathscr{A},$$

which define homeomorphisms, are both involved in a more essential way.

An n-dimensional manifold M, together with a covering

$$\mathscr{U} = [U_\alpha : \alpha \in \mathscr{A}],$$

of open sets in M, each homeomorphic to B_n, with homeomorphisms

$$\phi_\alpha : U_\alpha \to B_n,$$

is said to be a **differentiable manifold** if, for every $\alpha, \beta \in \mathscr{A}$, the mapping

$$\phi_\beta \circ \phi_\alpha^{-1} : \phi_\alpha(U_\alpha \cap U_\beta) \to \phi_\beta(U_\alpha \cap U_\beta)$$

is a regular differentiable mapping.

Remark Regularity follows from the tacit assumption that the inverse mappings are also differentiable, but we have chosen to emphasize the property.

If the mappings

$$\phi_\beta \circ \phi_\alpha^{-1}$$

are all of class C^k, the manifold is said to be of **class C^k**. If the mappings

$$\phi_\beta \circ \phi_\alpha^{-1}$$

are all of class C^∞, the manifold is said to be of **class C^∞**.

Remark We shall assume that all manifolds are of class C^∞ and refer to these as differentiable manifolds. However, we shall sometimes write C^∞ in parentheses, lest the reader forget.

The above definition merits discussion. For a given $\alpha, \beta \in \mathscr{A}$, the set $U_\alpha \cap U_\beta$ may be empty. Then

$$\phi_\alpha(U_\alpha \cap U_\beta) \qquad \text{and} \qquad \phi_\beta(U_\alpha \cap U_\beta)$$

are empty, and there is no issue. Suppose that $U_\alpha \cap U_\beta$ is not empty. Since $U_\alpha \cap U_\beta \subset U_\alpha$, we have $\phi_\alpha(U_\alpha \cap U_\beta) \subset B_n$. Similarly,

$$\phi_\beta(U_\alpha \cap U_\beta) \subset B_n.$$

On the other hand, the mapping

$$\phi_\beta \circ \phi_\alpha^{-1} : \phi_\alpha(U_\alpha \cap U_\beta) \to \phi_\beta(U_\alpha \cap U_\beta)$$

is a homeomorphism. Since this mapping is on an open set in euclidean n-space, it makes sense to discuss its differentiability properties.

The mapping ϕ_α defines coordinates in the open set U_α, $\alpha \in \mathscr{A}$. We shall, accordingly, refer to each U as a **coordinate neighborhood**. The coordinates themselves will be referred to as **local coordinates**. Thus, for every $p \in U_\alpha$, the local coordinates of p are the n-tuple, $(x_1, \cdots, x_n)$, where x_i is the ith coordinate of the point $\phi_\alpha(p)$, $i = 1, \cdots, n$.

If $p \in U_\alpha \cap U_\beta$, then p has local coordinates $(x_1, \cdots, x_n)$, determined by the mapping ϕ_α, and local coordinates $(y_1, \cdots, y_n)$, determined by the mapping ϕ_β. Moreover, in terms of these coordinates, the mapping

$$\phi_\beta \circ \phi_\alpha^{-1}$$

may be written as

$$y_1 = y_1(x_1, \cdots, x_n)$$
$$\cdots$$
$$y_n = y_n(x_1, \cdots, x_n).$$

It is important to observe that, while the coordinate neighborhoods U_α, $\alpha \in \mathscr{A}$, determine the differentiable structure of the manifold, there are in general other coordinate neighborhoods in a differentiable manifold. Indeed, the manifold may be determined as well by a completely different set V_β, $\beta \in \mathscr{B}$, with coordinate mappings ψ_β, $\beta \in \mathscr{B}$. We shall discuss this matter in the next section, when we consider diffeomorphisms between manifolds.

As an example of a differentiable manifold (of class C^∞), we consider the circle S, of radius 1. Consider the polar representation of this circle. Let $U_1 = S \sim \{(1, 0)\}$ and let $\phi_1: U_1 \to B_1$ be the mapping for which

$$\phi_1((1, \theta)) = \frac{\theta - \pi}{\pi}, \qquad 0 < \theta < 2\pi.$$

Let $U_2 = S \sim \{(1, \pi)\}$ and let $\phi_2: U_2 \to B_1$ be the mapping for which

$$\phi_2((1, \theta)) = \frac{\theta}{\pi}, \qquad -\pi < \theta < \pi.$$

We determine the mapping

$$\phi_2 \circ \phi_1^{-1},$$

which is a homeomorphism

$$\phi_2 \circ \phi_1^{-1}: B_1 \sim (\{0\}) \to B_1 \sim (\{0\}).$$

Indeed,

$$\phi_2 \circ \phi_1^{-1}(x) = \begin{cases} x + 1, & -1 < x < 0, \\ x - 1, & 0 < x < 1, \end{cases}$$

so that we have a differentiable (C^∞) manifold.

We make one more remark. The mappings ϕ_α give orientations to the sets U_α. We shall not discuss the orientation given to an individual U_α.

However, if $U_\alpha \cap U_\beta$ is nonempty, then U_α and U_β are given the same orientations if the Jacobian of the mapping

$$\phi_\beta \circ \phi_\alpha^{-1}$$

is positive. Otherwise, they are given opposite orientations. This suggests that we define a manifold to be **oriented** if, for every $\alpha, \beta \in \mathscr{A}$, and $U_\alpha \cap U_\beta$ nonempty,

$$\phi_\beta \circ \phi_\alpha^{-1}$$

has positive Jacobian.

4. DIFFERENTIABLE FUNCTIONS AND MAPPINGS

We define differentiable functions on a manifold. Let M be an n-dimensional differentiable manifold (of class C^∞), let $G \subset M$ be an open set, and let

$$f \colon G \to R$$

be a real function on G. Let U_α, $\alpha \in \mathscr{A}$, be coordinate neighborhoods which cover M and let ϕ_α be the corresponding mappings.

Let $p \in G \cap U_\alpha$. Then $f \circ \phi_\alpha^{-1}$ is defined on the open set $\phi_\alpha(G \cap U_\alpha) \subset B_n$. We say that f is **differentiable at** p if $f \circ \phi_\alpha^{-1}$ is differentiable at $\phi_\alpha(p)$; and that f is differentiable on $G \cap U_\alpha$ if $f \circ \phi_\alpha^{-1}$ is differentiable on $\phi_\alpha(G \cap U_\alpha)$. We say that f is **of class** C^∞ on $G \cap U_\alpha$, if $f \circ \phi_\alpha^{-1}$ is of class C^∞ on $\phi_\alpha(G \cap U_\alpha)$. We shall deal only with functions of class C^∞ on open sets and shall refer to these as differentiable functions.

In view of the fact that the mappings $\phi_\beta \circ \phi_\alpha^{-1}$ are all of class C^∞, it follows that the differentiability of f at p is independent of the choice of the U_α to which p belongs. Differentiability of f on G means differentiability on $U_\alpha \cap G$, for every $\alpha \in \mathscr{A}$.

For each $p \in M$, we shall consider the family of all functions, each of which is differentiable on an open set containing p. The open set varies with the function. We designate this class of functions as $\mathscr{F}_p$. We wish to add and multiply functions in $\mathscr{F}_p$. In view of the fact that the domains of functions are different, it is preferable to consider certain equivalence classes of functions in $\mathscr{F}_p$, rather than the functions themselves. These equivalence classes are called germs of differentiable (of class C^∞) functions at p.

The elements of $\mathscr{F}_p$ themselves are functions together with open sets, containing p, on which they are defined. Accordingly, the elements of $\mathscr{F}_p$ are pairs (f, G).

For $(f_1, G_1) \in \mathcal{F}_p$ and $(f_2, G_2) \in \mathcal{F}_p$, we shall say that (f_1, G_1) is equivalent to (f_2, G_2), if there is an open set H, such that $p \in H$ and $f_1(q) = f_2(q)$, for every $q \in H$. The reader may verify that the properties of an equivalence relation are obeyed by this relation, so that $\mathcal{F}_p$ is decomposed into disjoint equivalence classes. The equivalence classes are called **germs of differentiable (of class C^∞) functions** at p. The set of germs at p is designated as $\mathcal{G}_p$, and we use the symbols $\{f\}, \{g\}, \cdots$, to represent germs. Thus, each germ $\{f\}$ is composed of a set of pairs (f, G). Sometimes, we drop the braces and designate a germ simply as a function f. This will be done if there is no danger of confusion.

We define operations for germs. Let $\{f\}, \{g\} \in \mathcal{G}_p$ and let $(f, G) \in \{f\}$ and $(g, H) \in \{g\}$. Then $\{f\} + \{g\}$ is defined to be the germ to which $(f + g, G \cap H)$ belongs, and $\{f\} \cdot \{g\}$ is defined to be the germ to which $(fg, G \cap H)$ belongs. Moreover, $c\{f\}$ is defined to be the germ to which (cf, G) belongs. Under the first and third of these operations, $\mathcal{G}_p$ is a vector space, and with the second added, $\mathcal{G}_p$ is an algebra.

We also define differentiable mappings from one differentiable manifold to another. Let M be a differentiable manifold, covered by coordinate neighborhoods

$$U_\alpha, \alpha \in \mathcal{A},$$

and with differentiable structure given by mappings

$$\phi_\alpha \colon U_\alpha \to B_n;$$

and let N be a differentiable manifold, covered by coordinate neighborhoods

$$V_\beta, \beta \in \mathcal{B},$$

and with differentiable structure given by mappings

$$\psi_\beta \colon V_\beta \to B_m.$$

(Note that the dimensions of M and N may be different.) Let

$$f \colon M \to N$$

be continuous. We wish to define the differentiability of f in a neighborhood of $p \in M$. For this, let V_β be such that $f(p) \in V_\beta$. Since f is continuous, there is an open set G such that $p \in G$ and $f(G) \subset V_\beta$. There is an $\alpha \in \mathcal{A}$, such that $p \in U_\alpha$. We say that f is **differentiable** in a neighborhood of p, if there is an open set H, such that

$$p \in H, H \subset G \cap U_\alpha,$$

and the mapping

$$\psi_\beta \circ f \circ \phi_\alpha^{-1} \colon \phi_\alpha(H) \to \psi_\beta(V_\beta)$$

is differentiable (of class C^∞).

If the dimensions of M and N are the same, if f is bijective with inverse f^{-1}, and if f is differentiable in a neighborhood of each point of M and f^{-1} is differentiable in a neighborhood of each point of N, then f is said to be a **diffeomorphism.** Differentiable manifolds, with the same dimension, are then said to be **diffeomorphic** if a diffeomorphism exists between them.

Remark   It is known that there are differentiable manifolds which are homeomorphic but not diffeomorphic. It is also known that there are manifolds which do not admit a differentiable structure. The proofs of these facts are well outside of the scope of this book.

In particular, let M be a differentiable manifold, covered by co-ordinate neighborhoods

$$U_\alpha, \, \alpha \in \mathscr{A},$$

and with differentiable structure given by mappings

$$\phi_\alpha \colon U_\alpha \to B_n.$$

Consider a second differentiable structure in M, given by coordinate neighborhoods

$$V_\beta, \, \beta \in \mathscr{B},$$

which cover M, and mappings

$$\psi_\beta \colon V_\beta \to B_n.$$

Let us suppose that the identity mapping

$$i \colon M \to M$$

with the two differentiable structures, is a diffeomorphism.

We leave it to the reader to show that a necessary and sufficient condition for this is that M, with the coordinate neighborhoods U_α, $\alpha \in \mathscr{A}$ and $V_\beta, \, \beta \in \mathscr{B}$ and mappings $\phi_\alpha \colon U_\alpha \to B_n$ and $\psi_\beta \colon V_\beta \to B_n$, is still a differentiable manifold. This amounts to noting that the mappings

$$\phi_\alpha \circ \psi_\beta^{-1} \quad \text{and} \quad \psi_\beta \circ \phi_\alpha^{-1}$$

are differentiable for all $\alpha \in \mathscr{A}$ and $\beta \in \mathscr{B}$.

Moreover, for a differentiable manifold M, covered by coordinate neighborhoods

$$U_\alpha, \alpha \in \mathscr{A},$$

with mappings

$$\phi_\alpha \colon U_\alpha \to B_n,$$

an open set $U \subset M$, and homeomorphism,

$$\phi \colon U \to B_n$$

is called a **coordinate neighborhood** in the differentiable manifold M, if M is a differentiable manifold with coordinate neighborhoods

$$U_\alpha, \alpha \in \mathscr{A} \text{ and } U,$$

and mappings

$$\phi_\alpha \colon U_\alpha \to B_n \quad \text{and} \quad \phi \colon U \to B_n,$$

i.e., if

$$\phi_\alpha \circ \phi^{-1} \quad \text{and} \quad \phi \circ \phi_\alpha^{-1}$$

are differentiable for every $\alpha \in \mathscr{A}$.

We define a differentiable manifold M to be **orientable** if it has a covering

$$U_\alpha, \alpha \in \mathscr{A};$$

with mappings

$$\phi_\alpha \colon U_\alpha \to B_n,$$

for which all the diffeomorphisms

$$\phi_\beta \circ \phi_\alpha^{-1}$$

have positive Jacobians. We shall consider only orientable manifolds.

As a final definition, a set S in a manifold M is **compact** if every infinite subset of S has a limit point in S. We leave it for the reader to show that, for a manifold, this is equivalent to the property that if S is covered by a family of open sets, it is covered by a finite number of these open sets.

5. PARTITION OF UNITY

We consider now a special class of differentiable manifolds, which we call σ compact. We first obtain some preliminary facts.

Lemma 1

If B_n is the unit ball in R^n and
$$p \in B_n, \qquad p = (x_1, \cdots, x_n),$$
there is a diffeomorphism f of B_n onto itself such that $f(p) = \theta = (0, 0, \cdots, 0)$.

Proof

We merely note that the reader may obtain the desired mapping as a composition of n diffeomorphisms the first of which takes $(x_1, \cdots, x_n)$ into $(0, x_2, \cdots, x_n)$; the second of which takes $(0, x_2, \cdots, x_n)$ into $(0, 0, x_3, \cdots, x_n)$, etc. ∎

From this lemma, we may easily obtain

Lemma 2

If M is a differentiable manifold, $p \in M$, and G is an open set with $p \in G$, then there is a coordinate neighborhood U of p, $U \subset G$, with mapping $\phi \colon U \to B_n$, such that $\phi(p) = (0, \cdots, 0)$.

We leave the proof to the reader.

We say that a manifold M is σ **compact** if

$$M = \bigcup_{k=1}^{\infty} G_k,$$

where G_k is open, $\bar{G}_k$ is compact, and
$$\bar{G}_k \subset G_{k+1}, \qquad k = 1, 2, \cdots.$$

We prove

Theorem 1

If M is a σ-compact, differentiable manifold, with open sets G_k, $k = 1, 2, \cdots$ having the property of the definition, there is a countable set
$$U_i, \qquad i = 1, 2, \cdots$$
of coordinate neighborhoods, which covers M, with mappings
$$\phi_i \colon U_i \to B_n,$$
such that, for each $k = 1, 2, \cdots$, all but a finite number of the sets
$$G_k \cap U_i, \qquad i = 1, 2, \cdots$$
are empty, and the sets
$$V_i, \qquad i = 1, 2, \cdots$$
cover M, where
$$V_i = \phi_i^{-1}(A_n)$$
and
$$A_n = \{x \colon x \in B_n, |x| < \tfrac{1}{2}\}.$$

Proof

For each $k = 2, 3, \cdots$, let

$$S_k = \bar{G}_{k+1} \sim G_k.$$

For each $p \in S_k$, let U_p be a coordinate neighborhood of p, with mapping

$$\phi_p \colon U_p \to B_n,$$

and

$$\phi_p(p) = (0, \cdots, 0),$$

such that

$$U_p \subset G_{k+2} \sim \bar{G}_{k-1}.$$

Let

$$V_p = \phi_p^{-1}(A_n).$$

For $k = 1$, let

$$S_1 = \bar{G}_2.$$

For each $p \in S_1$, let U_p be a coordinate neighborhood containing p and with mapping

$$\phi_p \colon U_p \to B_n,$$

such that

$$\phi_p(p) = (0, \cdots, 0)$$

and $U_p \subset G_3$. Again, let

$$V_p = \phi_p^{-1}(A_n).$$

Each

$$S_k, \qquad k = 1, 2, \cdots$$

is compact. Hence, a finite number of the $V_p, p \in S_k$, cover S_k. Designate them by

$$V_{k,1}, \cdots, V_{k,n_k}.$$

The collection

$$V_{k,j}, \qquad k = 1, 2, \cdots; \quad j = 1, \cdots, n_k$$

has the required property.   ∎

It is essential for the definition of the integral of a differential form to introduce a partition of unity in a manifold. Theorem 1 allows us to do this for a σ-compact manifold. We must first consider a special function of a real variable. We show that the function f given by

$$f(x) = \begin{cases} e^{-1/x^2}, & x \neq 0, \\ 0, & x = 0, \end{cases}$$

is infinitely differentiable, and that all the derivatives are zero at $x = 0$.

Since

$$\lim_{x \to 0} \frac{1}{x} e^{-1/x^2} = 0,$$

we have

$$f'(x) = \begin{cases} \dfrac{2}{x^3} e^{-1/x^2}, & x \neq 0, \\ 0, & x = 0. \end{cases}$$

Suppose the nth derivative of f exists and is given by

$$f^n(x) = \begin{cases} p\left(\dfrac{1}{x}\right) e^{-1/x^2}, & x \neq 0, \\ 0, & x = 0, \end{cases}$$

where $p(1/x)$ is a polynomial in $1/x$. Then, since

$$\lim_{x \to 0} \frac{1}{x} p\left(\frac{1}{x}\right) e^{-1/x^2} = 0,$$

the $(n + 1)$st derivative of f also exists and is given by

$$f^{n+1}(x) = \begin{cases} q\left(\dfrac{1}{x}\right) e^{-1/x^2}, & x \neq 0, \\ 0, & x = 0, \end{cases}$$

where $q(1/x)$ is a polynomial in $1/x$. This induction argument proves the assertion.

We leave it to the reader to see that the function f given by

$$f(x) = \begin{cases} \exp\left\{-\left[\dfrac{1}{x + \frac{1}{2}} + \dfrac{1}{\frac{1}{2} - x}\right]^2\right\}, & |x| < \frac{1}{2}, \\ 0, & |x| \geq \frac{1}{2}, \end{cases}$$

is infinitely differentiable, positive on $(-1/2, 1/2)$, and zero everywhere else.

The function $f\colon R^n \to R$, given by

$$f(x) = \begin{cases} \exp\left\{-\left[\dfrac{1}{|x| + \frac{1}{2}} + \dfrac{1}{\frac{1}{2} - |x|}\right]\right\}^2, & |x| < \frac{1}{2}, \\ 0, & |x| \geq \frac{1}{2}, \end{cases}$$

is then infinitely differentiable, positive on A_n, and zero everywhere else.

Now, let M be a σ-compact differentiable manifold with coordinate neighborhoods U_i, mappings

$$\phi_i: U_i \to B_n,$$

and open sets

$$V_i, \qquad i = 1, 2, \cdots,$$

as described in Theorem 1. For each $i = 1, 2, \cdots$, consider the function

$$\psi_i: M \to R$$

defined by

$$\psi_i(p) = \begin{cases} f \circ \phi_i(p), & p \in U_i, \\ 0, & p \notin U_i, \end{cases}$$

where f is the above real function on R^n.

The function ψ_i is then infinitely differentiable on M, positive on V_i, and zero everywhere else.

It then follows, by Theorem 1, that for every $p \in M$,

$$\psi_i(p) = 0,$$

except for a finite number of values of i. Thus the function

$$\sum_{i=1}^{\infty} \psi_i$$

is positive and finite everywhere.

Now, for every $i = 1, 2, \cdots$, let

$$g_i = \frac{\psi_i}{\displaystyle\sum_{i=1}^{\infty} \psi_i}.$$

Then

$$g_i(p) > 0, \qquad p \in V_i,$$
$$= 0, \qquad p \notin V_i,$$

and

$$\sum_{i=1}^{\infty} g_i(p) = 1, \qquad p \in M.$$

We note that, for each $i = 1, 2, \cdots$, there is a coordinate neighborhood U_i such that g_i vanishes on the complement of U_i.

A sequence of functions, with the properties of the above functions g_i, $i = 1, 2, \cdots$, is called a **partition of unity**.

6. TANGENT SPACE

Let M be a differentiable manifold and let $p \in M$. We define the tangent space to the manifold M at the point p. Let $\mathscr{G}_p$ be the set of germs of differentiable real functions at p. In the notation $\{f\}$ for a germ, we shall drop the braces, as there is no danger of confusion.

$\mathscr{G}_p$ is a vector space. Indeed, with the product as defined in Section 3, it is an algebra, i.e., it is an algebra under the operations $f + g$, fg, and cf. Since $\mathscr{G}_p$ is a vector space, we may consider linear functionals on $\mathscr{G}_p$. These are mappings

$$\phi\colon \mathscr{G}_p \to R$$

which are such that

(a) $\qquad \phi(f + g) = \phi(f) + \phi(g), \qquad$ for all $f, g \in \mathscr{G}_p$,

(b) $\qquad \phi(cf) = c\phi(f), \qquad$ for all $f \in \mathscr{G}_p$, $c \in R$.

We consider only those linear ϕ which also satisfy

(c) $\qquad \phi(fg) = g(p)\phi(f) + f(p)\phi(g), \qquad$ for all $f, g \in \mathscr{G}_p$.

We let $\mathscr{T}_p$ be the set of all ϕ which satisfy (a), (b), and (c).

Each $\phi \in \mathscr{T}_p$ is called a **tangent vector** to the manifold M at the point p, and $\mathscr{T}_p$ is called the **tangent space** to M at p.

It is clear that $\mathscr{T}_p$ is a vector space, i.e., if

$$\phi_1, \phi_2 \in \mathscr{T}_p,$$

then

$$\phi_1 + \phi_2 \in \mathscr{T}_p,$$

and if

$$\phi \in \mathscr{T}_p, \quad c \in R,$$

then

$$c\phi \in \mathscr{T}_p.$$

We show that $\mathscr{T}_p$ is n-dimensional, i.e., the same as the dimensionality of M. In doing this, we note that each $\phi \in \mathscr{T}_p$ is a functional, satisfying (a), (b), and (c) on $\mathscr{F}_p$ as well as on $\mathscr{G}_p$, which is constant on those subsets of $\mathscr{F}_p$ which are elements of $\mathscr{G}_p$. We may thus consider ϕ as acting on functions in neighborhoods of p.

Let $\phi \in \mathscr{T}_p$. Consider the constant function $f = 1$. By (c),

$$\phi(1) = \phi(1 \cdot 1) = \phi(1) + \phi(1),$$

so that

$$\phi(1) = 0.$$

It follows, by (b), that for all constant functions c,

$$\phi(c) = c\phi(1) = 0.$$

Let $f \in \mathscr{F}_p$ and let U be a coordinate neighborhood of p, such that f is defined on U. Let $(x_1, \cdots, x_n)$ be the coordinates in U and let ψ be the coordinate mapping. Designate $f \circ \psi^{-1}$ by f and $\psi(p)$ by $x^0 = (x_1^0, \cdots, x_n^0)$.

Since f is infinitely differentiable,

$$f(x) = f(x^0) + \sum_{i=1}^{n} (x_i - x_i^0)\frac{\partial f}{\partial x_i}(x^0) + \sum_{i,j=1}^{n} (x_i - x_i^0)(x_j - x_j^0)f_{ij}(x),$$

where each f_{ij} is infinitely differentiable.

Now,

$$\phi(f) = \phi(f(x^0)) + \sum_{i=1}^{n} \frac{\partial f}{\partial x_i}(x^0)\phi(x_i - x_i^0)$$

$$+ \sum_{i,j=1}^{n} \phi[(x_i - x_i^0)(x_j - x_j^0)f_{ij}(x)].$$

By an easy application of (c),

$$\sum_{i,j=1}^{n} \phi[(x_i - x_i^0)(x_j - x_j^0)f_{ij}(x)] = 0.$$

Moreover,

$$\phi(f(x^0)) = 0,$$

and

$$\sum_{i=1}^{n} \frac{\partial f}{\partial x_i}(x^0)\phi(x_i^0) = 0.$$

Hence

(*) $$\phi(f) = \sum_{i=1}^{n} \frac{\partial f}{\partial x_i}(x^0)\phi(x_i).$$

The formula (*) is suggestive of the distinction of special tangent vectors at p. These will be designated as

$$\frac{\partial}{\partial x_1}\bigg|_p, \cdots, \frac{\partial}{\partial x_n}\bigg|_p,$$

or just as

$$\frac{\partial}{\partial x_1}, \cdots, \frac{\partial}{\partial x_n},$$

where the evaluation at p is understood.

The vector $\partial/\partial x_i$ is defined to be that linear functional on $\mathscr{F}_p$ for which

$$\frac{\partial}{\partial x_i}(f) = \frac{\partial f}{\partial x_i}(x^0).$$

This functional assumes the same value for f and g belonging to the same element of $\mathscr{G}_p$; hence $\partial/\partial x_i$ is defined on $\mathscr{G}_p$.

Moreover, $\partial/\partial x_i$ is a tangent vector, since

(a) $$\frac{\partial(f + g)}{\partial x_i}(x^0) = \frac{\partial f}{\partial x_i}(x^0) + \frac{\partial g}{\partial x_i}(x^0),$$

(b) $$\frac{\partial cf}{\partial x_i}(x^0) = c\frac{\partial f}{\partial x_i}(x^0),$$

(c) $$\frac{\partial(fg)}{\partial x_i}(x^0) = f(x^0)\frac{\partial g}{\partial x_i}(x^0) + g(x^0)\frac{\partial f}{\partial x_i}(x^0).$$

By (*), we have, for every $\phi \in \mathscr{T}_p$,

$$\phi = \sum_{i=1}^{n} \phi(x_i)\frac{\partial}{\partial x_i},$$

so that every

$$\phi \in \mathscr{T}_p$$

is a linear combination of

$$\frac{\partial}{\partial x_1}, \cdots, \frac{\partial}{\partial x_n}.$$

Moreover, these n-vectors are linearly independent as the reader may verify by considering their values at the coordinate functions $x_1, \cdots, x_n$. We have thus proved

Theorem 2

The tangent space $\mathscr{T}_p$ to M at p is n-dimensional.

Of course, if a different coordinate neighborhood of p were considered, a different basis for $\mathscr{T}_p$ would be obtained.

Finally, it is of interest to note that $\mathscr{T}_p$ is the dual of an n-dimensional space, obtained from $\mathscr{G}_p$ as a quotient space. Let U be a coordinate neighborhood of p, with coordinate mapping ϕ. For each $f \in \mathscr{F}_p$, consider the differential of the function $f \circ \phi^{-1}$ at $\phi(p)$. If f and g belong to the same germ, they have the same differential. Thus this differential is a function on $\mathscr{G}_p$. Let $\mathcal{O}_p \subset \mathscr{G}_p$ be that set for which the differential is zero. Then $f \in \mathcal{O}_p$ if and only if

$$\frac{\partial f}{\partial x_i} = 0, \qquad i = 1, \cdots, n.$$

In other words, $f \in \mathcal{O}_p$ if and only if $\phi(f) = 0$ for every $\phi \in \mathscr{T}_p$.

We leave the remaining details of the proof that $\mathscr{G}_p/\mathcal{O}_p$ is n-dimensional and that $\mathscr{T}_p$ is its dual to the reader.

7. SPACE OF DIFFERENTIALS

Since $\mathscr{T}_p$ is an n-dimensional vector space, its dual space $\mathscr{T}_p^*$ is also n-dimensional. Given local coordinates $(x_1, \cdots, x_n)$ in a neighborhood of p, we have seen that a basis for $\mathscr{T}_p$ is given by the vectors

$$\frac{\partial}{\partial x_1}, \cdots, \frac{\partial}{\partial x_n},$$

where the evaluation is understood to be at p.

Since $\mathscr{T}_p$ is the dual of the vector space $\mathscr{G}_p/\mathcal{O}_p$, we have seen in Chapter I that there is a natural isomorphism between $\mathscr{T}_p^*$ and $\mathscr{G}_p/\mathcal{O}_p$. Let us designate this isomorphism by

$$\Phi: \mathscr{G}_p/\mathcal{O}_p \to \mathscr{T}_p^*.$$

Let us consider the functions $x_1, \cdots, x_n$ in $\mathscr{F}_p$. Each of these functions represents an element of $\mathscr{G}_p/\mathcal{O}_p$, which we again designate as x_i, $i = 1, \cdots, n$. Since

$$\frac{\partial x_i}{\partial x_j} = \delta_{ij},$$

it is clear that

$$\Phi(x_1), \cdots, \Phi(x_n)$$

is the dual basis in $\mathscr{T}_p^*$ to the basis

$$\frac{\partial}{\partial x_1}, \cdots, \frac{\partial}{\partial x_n}$$

in $\mathscr{T}_p$.

In the same way, for every $f \in \mathscr{F}_p$, the

$$f \in \mathscr{G}_p/\mathcal{O}_p,$$

which it defines, determines the element

$$\Phi(f) \in \mathscr{T}_p^*$$

whose values at the basis elements of $\mathcal{T}_p$ are

$$\frac{\partial f}{\partial x_1}(p), \cdots, \frac{\partial f}{\partial x_n}(p).$$

The element

$$\Phi(f) \in \mathcal{T}_p^*,$$

which corresponds to f, is called the differential of f and will be designated as df. In particular, for the coordinate functions x_i, $i = 1, \cdots, n$, the corresponding $\Phi(x_i)$ are the differentials dx_i, $i = 1, \cdots, n$.

Remark In Chapter II, in the case of the manifold R^n, the differential of a function at a point was also introduced as a linear functional; in that case, presumably on R^n itself. In reality, the functional then was also on the tangent space at the point in question. The fact that euclidean space is also a vector space makes an identification of the two spaces natural, and it indicated, for pedagogical reasons, that a distinction which might seem to be a quibble had better not be made.

The differentials $dx_1, \cdots, dx_n$ form a basis in $\mathcal{T}_p^*$. As noted above, they are the dual basis to the basis

$$\frac{\partial}{\partial x_1}, \cdots, \frac{\partial}{\partial x_n}, \quad \text{in } \mathcal{T}_p.$$

Let $f \in \mathcal{F}_p$. Let us evaluate df at the vectors

$$\frac{\partial}{\partial x_1}, \cdots, \frac{\partial}{\partial x_n}.$$

This evaluation is simply

$$(df)\left(\frac{\partial}{\partial x_i}\right) = \frac{\partial f}{\partial x_i}(p).$$

In view of the fact that

$$(dx_j)\left(\frac{\partial}{\partial x_i}\right) = \delta_{ji},$$

we get

$$df = \frac{\partial f}{\partial x_1}(p)\, dx_1 + \cdots + \frac{\partial f}{\partial x_n}(p)\, dx_n.$$

Now, suppose p lies in coordinate neighborhoods given by coordinates $(x_1, \cdots, x_n)$ and $(y_1, \cdots, y_n)$, respectively. As a special case, we then have

$$dy_i = \frac{\partial y_i}{\partial x_1}(p)\, dx_i + \cdots + \frac{\partial y_i}{\partial x_n}(p)\, dx_n, \qquad i = 1, 2, \cdots, n.$$

We now give the main definition. A **differential form** (1-form) on a manifold M is a mapping which associates, with each $p \in M$, an element of $\mathscr{T}_p^*$.

Suppose ω is a differential form on M, let U be a coordinate neighborhood on M, with coordinates $(x_1, \cdots, x_n)$. Since the differentials $dx_1, \cdots, dx_n$ form a basis for $\mathscr{T}_p^*$ at each $p \in U$, ω is given by

$$\omega(p) = a_1(p)\, dx_1 + \cdots + a_n(p)\, dx_n$$

on U.

If V is another coordinate neighborhood on M, with coordinates $(y_1, \cdots, y_n)$, then

$$\omega(p) = b_1(p)\, dy_1 + \cdots + b_n(p)\, dy_n.$$

On $U \cap V$, expressions relating the a_i and b_i may be obtained in view of the relations

$$dy_i = \sum_{j=1}^{n} \frac{\partial y_i}{\partial x_j}\, dx_j.$$

These are

$$a_j = \sum_{i=1}^{n} b_i \frac{\partial y_i}{\partial x_j} \quad \text{or} \quad b_j = \sum_{i=1}^{n} a_i \frac{\partial x_i}{\partial y_j}.$$

8. GRASSMANN ALGEBRA

The differential forms defined in the preceding section are 1-forms. Although they are defined on an n-dimensional manifold M, they are the objects for which integrals on one-dimensional manifolds (imbedded in M) are defined. Indeed, there are differential forms of order p, for every $1 \leq p \leq n$. These so-called p-forms may be obtained from the 1-forms.

In order to see how this is done, it is desirable to discuss the Grassmann algebra associated with a vector space. We recall that an **algebra** A has three operations, addition, multiplication by reals, and multiplication, such that A is a vector space under the first two operations and a ring under the first and third. An algebra may or may not have a multiplicative identity, and multiplication may or may not be commutative. A set $S \subset A$ is said to generate A if there is no proper subalgebra of A which contains S.

Let V be a vector space of dimension n. We show that there is an algebra $G(V)$, containing V and the real numbers, such that:

(a) 1 and V generate $G(V)$.

(b) 1 is the multiplicative identity for $G(V)$.

(c) Addition, multiplication by reals, and addition and multiplication of reals by reals is the same in $G(V)$ as it is in V and R, when applied to the elements of $G(V)$ which belong to V and R.

(d) If multiplication in $G(V)$ is denoted by $x \wedge y$, then for all $x, y \in V$, we have

$$x \wedge y = -y \wedge x.$$

(e) There is a basis $e_1, \cdots, e_n$ in V such that

$$e_1 \wedge \cdots \wedge e_n \neq 0.$$

We construct $G(V)$ from V by considering, for a given basis $e_1, \cdots, e_n$ of V, the nonzero elements

$$1, \quad e_i, \quad i = 1, \cdots, n,$$

$$e_{i_1} \wedge e_{i_2}, \quad i_1 < i_2, \quad 1 \leq i_1, i_2 \leq n,$$

$$e_{i_1} \wedge e_{i_2} \wedge e_{i_3}, \quad i_1 < i_2 < i_3, \quad 1 \leq i_1, i_2, i_3 \leq n,$$

$$\cdots$$

$$e_{i_1} \wedge e_{i_2} \wedge \cdots \wedge e_{i_p}, \quad i_1 < i_2 < \cdots < i_p, \quad 1 \leq i_1, \cdots, i_p \leq n,$$

$$\cdots$$

$$e_1 \wedge \cdots \wedge e_n$$

The number of elements in the above set B is

$$\binom{n}{0} + \binom{n}{1} + \binom{n}{2} + \cdots + \binom{n}{n} = (1 + 1)^n = 2^n.$$

We consider the vector space $G(V)$ of finite linear combinations, with real coefficients, of elements of B. Multiplication in $G(V)$ may be defined by defining

$$(e_{i_1} \wedge \cdots \wedge e_{i_p}) \wedge (e_{j_1} \wedge \cdots \wedge e_{j_q}) = e_{i_1} \wedge \cdots \wedge e_{i_p} \wedge e_{j_1} \wedge \cdots \wedge e_{j_q},$$

subject to the rule

$$e_i \wedge e_j = -e_j \wedge e_i$$

and using the distributive law.

We then have—

(a) For each $x \in V$, $x \wedge x = 0$. This is easy and is left to the reader.
(b) The set B is linearly independent so that $G(V)$ is 2^n-dimensional.

Suppose we have a finite linear combination of the elements of B which is equal to 0. There will be a p such that

$$\sum_{i_1 < i_2 < \cdots < i_p} a_{i_1 \ldots i_p} e_{i_1} \wedge e_{i_2} \wedge \cdots \wedge e_{i_p} + \text{terms with more factors} = 0.$$

Consider any $i_1 < \cdots < i_p$. This is a subset of $1, 2, \cdots, n$. Let $j_1, \cdots, j_{n-p}$ be the complement of $i_1, \cdots, i_p$ in $1, 2, \cdots, n$. If we multiply the left-hand side above by

$$e_{j_1} \wedge \cdots \wedge e_{j_{n-p}},$$

all terms except one have a repeated e_i, and so they are 0. We are left with

$$\pm a_{i_1 \ldots i_p} e_1 \wedge \cdots \wedge e_n = 0.$$

Since

$$e_1 \wedge \cdots \wedge e_n \neq 0,$$

it follows that

$$a_{i_1 \ldots i_p} = 0. \quad \blacksquare$$

This proves the assertion. Here and elsewhere, the case $p = 0$ may be treated separately, in a similar way.
(c) $G(V)$ is an algebra generated by 1 and V. (It is called the **Grassmann algebra** over V.)

We leave the proof to the reader.

If $X_1, \cdots, X_k$ are vector spaces, we define the direct sum

$$X_1 \dotplus X_2 \dotplus \cdots \dotplus X_k$$

to be the vector space whose elements are k-tuples

$$x = (x_1, \cdots, x_k),$$

where

$$x_i \in X_i, \qquad i = 1, \cdots, k,$$

and addition and multiplication by reals are defined by

$$(x_1, \cdots, x_k) + (y_1, \cdots, y_k) = (x_1 + y_1, \cdots, x_k + y_k)$$

and

$$a(x_1, \cdots, x_k) = (ax_1, \cdots, ax_k),$$

respectively.

We consider special vector subspaces of $G(V)$. Define

V^0 to be the subspace generated by 1,

V^1 to be the subspace generated by e_i, $i = 1, \cdots, n$,

V^2 to be the subspace generated by

$$e_{i_1} \wedge e_{i_2}, \qquad i_1 < i_2, 1 \leq i_1, i_2 \leq n,$$

V^p to be the subspace generated by

$$e_{i_1} \wedge \cdots \wedge e_{i_p}, \qquad i_1 < \cdots < i_p, 1 \leq i_1, \cdots, i_p \leq n,$$

V^n to be the subspace generated by $e_1 \wedge \cdots \wedge e_n$.

Then

$$G(V) = V^0 \dotplus V^1 \dotplus V^2 \dotplus \cdots \dotplus V^n.$$

The arbitrary element of $V^p, p > 0$, is of the form

$$\sum_{i_1 < \cdots < i_p} a_{i_1 \cdots i_p} e_{i_1} \wedge \cdots \wedge e_{i_p}.$$

These members of $G(V)$ are said to be **homogeneous of order p.**

Let x be homogeneous of order p and y homogeneous of order q. We prove

$$x \wedge y = (-1)^{pq} y \wedge x.$$

Proof

It suffices, by linearity, to suppose

$$x = e_{i_1} \wedge \cdots \wedge e_{i_p} \qquad \text{and} \qquad y = e_{j_1} \wedge \cdots \wedge e_{j_q}.$$

Then

$$\begin{aligned}
x \wedge y &= (e_{i_1} \wedge \cdots \wedge e_{i_p}) \wedge (e_{j_1} \wedge \cdots \wedge e_{j_q}) \\
&= (-1)^p e_{j_1} \wedge (e_{i_1} \wedge \cdots \wedge e_{i_p}) \wedge (e_{j_2} \wedge \cdots \wedge e_{j_q}) \\
&= (-1)^{pq} (e_{j_1} \wedge \cdots \wedge e_{j_q}) \wedge (e_{i_1} \wedge \cdots \wedge e_{i_p}) \\
&= (-1)^{pq} y \wedge x. \qquad \blacksquare
\end{aligned}$$

Now, let

$$y_1, \cdots, y_p, \qquad 1 \leq p \leq n$$

be elements of V. Then, since $e_1, \cdots, e_n$ is a basis in V, we have

$$y_1 = \sum_{i=1}^{n} a_{1i} e_i$$

$$y_2 = \sum_{i=1}^{n} a_{2i} e_i$$

$$\cdots$$

$$y_p = \sum_{i=1}^{n} a_{pi} e_i.$$

We have

Proposition 1

$$y_1 \wedge y_2 \wedge \cdots \wedge y_p = \sum_{i_1 < \cdots < i_p} \begin{vmatrix} a_{1i_1} & \cdots & a_{1i_p} \\ & \cdots & \\ a_{pi_1} & \cdots & a_{pi_p} \end{vmatrix} e_{i_1} \wedge \cdots \wedge e_{i_p}.$$

We shall prove this proposition for the special case $p = 2$, $n = 3$. Then

$$y_1 = a_{11}e_1 + a_{12}e_2 + a_{13}e_3,$$
$$y_2 = a_{21}e_1 + a_{22}e_2 + a_{23}e_3.$$

Then
$$y_1 \wedge y_2 = (a_{11}e_1 + a_{12}e_2 + a_{13}e_3) \wedge (a_{21}e_1 + a_{22}e_2 + a_{23}e_3)$$
$$= (a_{11}a_{22} - a_{12}a_{21})e_1 \wedge e_2 + (a_{11}a_{23} - a_{13}a_{21})e_1 \wedge e_3$$
$$+ (a_{12}a_{23} - a_{13}a_{22})e_2 \wedge e_3$$
$$= \begin{vmatrix} a_{11} & a_{12} \\ a_{21} & a_{22} \end{vmatrix} e_1 \wedge e_2 + \begin{vmatrix} a_{11} & a_{13} \\ a_{21} & a_{23} \end{vmatrix} e_1 \wedge e_3 + \begin{vmatrix} a_{12} & a_{13} \\ a_{22} & a_{23} \end{vmatrix} e_2 \wedge e_3. \quad \blacksquare$$

Corollary

For the special case $p = n$, we have

$$y_1 \wedge y_2 \wedge \cdots \wedge y_n = \begin{vmatrix} a_{11} & \cdots & a_{1n} \\ & \cdots & \\ a_{n1} & \cdots & a_{nn} \end{vmatrix} e_1 \wedge \cdots \wedge e_n.$$

9. DIFFERENTIAL FORMS

We now return to the manifold M. Let $p \in M$ and let $V_p = \mathscr{T}_p^*$. Then with coordinates $(x_1, \cdots, x_n)$, V_p has a basis

$$dx_1, dx_2, \cdots, dx_n.$$

Let us consider the Grassmann algebra

$$G(V_p) = V_p^0 + V_p^1 + \cdots + V_p^n.$$

V_p^0 is the one-dimensional vector space of real numbers. V_p^1 is the vector space of elements

$$\sum_{i=1}^n a_i \, dx_i.$$

V_p^2 is the vector space of elements

$$\sum_{i_1 < i_2} a_{i_1 i_2} \, dx_{i_1} \wedge dx_{i_2}.$$

V_p^k is the vector space of elements

$$\sum_{i_1 < i_2 < \cdots < i_k} a_{i_1 \cdots i_k} \, dx_{i_1} \wedge \cdots \wedge dx_{i_k}.$$

The dimensionality of this space is $\binom{n}{k}$, the number of combinations of n things taken k at a time.

V_p^n is the one-dimensional vector space of elements

$$a \, dx_1 \wedge \cdots \wedge dx_n.$$

A **k-form** ω on M is a mapping on M whose value, for each $p \in M$, is an element of V_p^k.

Given a coordinate neighborhood U in M, a k-form ω may be expressed as

$$\omega(p) = \sum_{i_1 < \cdots < i_k} a_{i_1 \cdots i_k}(x_1, \cdots, x_n) \, dx_{i_1} \wedge \cdots \wedge dx_{i_k}.$$

A k-form is said to be **differentiable** (of class C^∞) if in each coordinate neighborhood the coefficient functions

$$a_{i_1 \cdots i_k}(x_1, \cdots, x_n)$$

are of class C^∞. We shall assume that all forms are differentiable.

For each $k = 1, \cdots, n$, let V^k be the set of all k-forms on M. Then V^k is easily seen to be a vector space. Moreover,

$$G = V^0 + V^1 + \cdots + V^n$$

is the algebra of forms on M, not necessarily homogeneous. Locally, an element of G may be written as

$$a(x_1, \cdots, x_n) + \sum_{i=1}^n a_i(x_1, \cdots, x_n) \, dx_i + \sum_{i_1 < i_2} a_{i_1 i_2}(x_1, \cdots, x_n) \, dx_{i_1} \wedge dx_{i_2}$$

$$+ \cdots + \sum_{i_1 < \cdots < i_k} a_{i_1 \cdots i_k}(x_1, \cdots, x_n) \, dx_{i_1} \wedge \cdots \wedge dx_{i_k} + \cdots +$$

$$a_{1 \cdots n}(x_1, \cdots, x_n) \cdot dx_1 \wedge \cdots \wedge dx_n.$$

We define an operator d, called **exterior differentiation** on the differential forms. For each $\omega \in V^k$, we define $d\omega$ as follows:

In a coordinate neighborhood with coordinates $(x_1, \cdots, x_n)$, if

$$\omega = a(x_1, \cdots, x_n)\, dx_{i_1} \wedge \cdots \wedge dx_{i_k},$$

then

$$d\omega = da \wedge dx_{i_1} \wedge \cdots \wedge dx_{i_k},$$

where

$$da = \sum_{i=1}^{n} \frac{\partial a}{\partial x_i}\, dx_i.$$

Then

$$d\omega = \sum_{i=1}^{n} \left(\frac{\partial a}{\partial x_i}\, dx_i \right) \wedge dx_{i_1} \wedge \cdots \wedge dx_{i_k}.$$

For an arbitrary k-form

$$\omega = \sum_{i_1 < \cdots < i_k} a_{i_1 \cdots i_k}\, dx_{i_1} \wedge \cdots \wedge dx_{i_k},$$

$d\omega$ is defined by linearity as

$$d\omega = \sum_{i_1 < \cdots < i_k} \sum_{i=1}^{n} \left(\frac{\partial a_{i_1 \cdots i_k}}{\partial x_i}\, dx_i \right) \wedge dx_{i_1} \wedge \cdots \wedge dx_{i_k}.$$

It is clear that d is a vector-space homomorphism

$$d: V^k \to V^{k+1}, \qquad k = 0, 1, \cdots, n - 1.$$

Clearly, $d\omega = 0$ for every n-form ω.

(a) A form ω is said to be **closed** if $d\omega = 0$.

(b) A form ω is said to be **exact** if there is a form θ such that $d\theta = \omega$.

Theorem 3

   Every exact form is closed; i.e., for every form ω, we have $d\, d\omega = 0$.

Proof

   Since d is linear, we need only consider a monomial. Let

$$\omega = a(x_1, \cdots, x_n)\, dx_{i_1} \wedge \cdots \wedge dx_{i_k},$$

in a coordinate neighborhood. Then

$$d\omega = \sum_{i=1}^{n} \frac{\partial a}{\partial x_i} dx_i \wedge dx_{i_1} \wedge \cdots \wedge dx_{i_k}$$

and

$$d\,d\omega = \sum_{i=1}^{n} d\left(\frac{\partial a}{\partial x_i}\right) \wedge dx_i \wedge dx_{i_1} \wedge \cdots \wedge dx_{i_k}$$

$$= \sum_{i,j=1}^{n} \frac{\partial^2 a}{\partial x_i\, \partial x_j} dx_j \wedge dx_i \wedge dx_{i_1} \wedge \cdots \wedge dx_{i_k}$$

$$= \sum_{i<j} \frac{\partial^2 a}{\partial x_i\, \partial x_j} (dx_i \wedge dx_j + dx_j \wedge dx_i) \wedge dx_{i_1} \wedge \cdots \wedge dx_{i_k}$$

$$= 0. \quad\blacksquare$$

The converse of Theorem 3 is false, except for very special manifolds, and we shall discuss this point briefly at the end of this chapter.

Now, we obtain an expression for the exterior derivative of a product.

Proposition 2

Let ω be a k-form and θ an l-form. Then

$$d(\omega \wedge \theta) = d\omega \wedge \theta + (-1)^k \omega \wedge d\theta.$$

Proof

It is only necessary to consider monomials

$$\omega = a\, dx_{i_1} \wedge \cdots \wedge dx_{i_k}$$

and

$$\theta = b\, dx_{j_1} \wedge \cdots \wedge dx_{j_l}.$$

Then

$$d(\omega \wedge \theta) = d(ab\, dx_{i_1} \wedge \cdots \wedge dx_{i_k} \wedge dx_{j_1} \wedge \cdots \wedge dx_{j_l})$$

$$= d(ab) \wedge dx_i \wedge \cdots \wedge dx_{i_k} \wedge dx_{j_1} \wedge \cdots \wedge dx_{j_l}$$

$$= \sum_{i=1}^{n} \left(a \frac{\partial b}{\partial x_i} + b \frac{\partial a}{\partial x_i}\right) dx_i \wedge dx_{i_1} \wedge \cdots \wedge dx_{i_k} \wedge dx_{j_1} \wedge \cdots \wedge dx_{j_l}$$

$$= \sum_{i=1}^{n} b \frac{\partial a}{\partial x_i} dx_i \wedge dx_{i_1} \wedge \cdots \wedge dx_{i_k} \wedge dx_{j_1} \wedge \cdots \wedge dx_{j_l}$$

$$+ (-1)^k \sum_{i=1}^{n} a \frac{\partial b}{\partial x_i} dx_{i_1} \wedge \cdots \wedge dx_{i_k} \wedge dx_i \wedge dx_{j_1} \wedge \cdots \wedge dx_{j_l}$$

$$= d\omega \wedge \theta + (-1)^k \omega \wedge d\theta. \quad\blacksquare$$

As a simple example, we have the euclidean 3-space R^3. Here we have one coordinate system (x, y, z). We may exhibit all differential forms.

The 0-forms are

$$\omega_0 = a(x, y, z).$$

The 1-forms are

$$\omega_1 = a \, dx + b \, dy + c \, dz.$$

The 2-forms are

$$\omega_2 = a \, dx \wedge dy + b \, dy \wedge dz + c \, dx \wedge dz.$$

The 3-forms are

$$\omega_3 = a \, dx \wedge dy \wedge dz.$$

Now,

$$d\omega_0 = da = \frac{\partial a}{\partial x} dx + \frac{\partial a}{\partial y} dy + \frac{\partial a}{\partial z} dz,$$

$$d\omega_1 = d(a \, dx + b \, dy + c \, dz)$$

$$= \left(\frac{\partial a}{\partial x} dx + \frac{\partial a}{\partial y} dy + \frac{\partial a}{\partial z} dz \right) \wedge dx + \left(\frac{\partial b}{\partial x} dx + \frac{\partial b}{\partial y} dy + \frac{\partial b}{\partial z} dz \right) \wedge dy$$

$$+ \left(\frac{\partial c}{\partial x} dx + \frac{\partial c}{\partial y} dy + \frac{\partial c}{\partial z} dz \right) \wedge dz$$

$$= \left(\frac{\partial b}{\partial x} - \frac{\partial a}{\partial y} \right) dx \wedge dy + \left(\frac{\partial c}{\partial y} - \frac{\partial b}{\partial z} \right) dy \wedge dz$$

$$+ \left(\frac{\partial c}{\partial x} - \frac{\partial a}{\partial z} \right) dx \wedge dz,$$

$$d\omega_2 = d(a \, dx \wedge dy + b \, dy \wedge dz + c \, dx \wedge dz)$$

$$= \frac{\partial a}{\partial z} dz \wedge dx \wedge dy + \frac{\partial b}{\partial x} dx \wedge dy \wedge dz + \frac{\partial c}{\partial y} dy \wedge dx \wedge dz$$

$$= \left(\frac{\partial a}{\partial z} + \frac{\partial b}{\partial x} - \frac{\partial c}{\partial y} \right) dx \wedge dy \wedge dz,$$

$$d\omega_3 = 0.$$

10. INTEGRAL OF A FORM; STOKES' THEOREM

We now define the integral of an n-form on an n-dimensional manifold M. For the sake of simplicity, we shall assume that M is compact and oriented.

Let $U_1, \cdots, U_m$ be a covering of M by coordinate neighborhoods and let $F_1, \cdots, F_m$ be a corresponding partition of unity. Then

$$F_i(p) \geq 0, \qquad p \in M, i = 1, \cdots, m,$$

$$F_i(p) = 0, \qquad p \notin U_i, i = 1, \cdots, m,$$

$$\sum_{i=1}^{m} F_i(p) = 1, \qquad p \in M,$$

and F_i is infinitely differentiable, $i = 1, \cdots, m$.

Let ω be a differential n-form on M. If the local coordinates in U_i are $(x_1, \cdots, x_n)$, then

$$\omega = a_i(x_1, \cdots, x_n) \, dx_1 \wedge \cdots \wedge dx_n, \qquad \text{on } U_i.$$

If the local coordinates in U_j are $(y_1, \cdots, y_n)$, then

$$\omega = a_j(y_1, \cdots, y_n) \, dy_1 \wedge \cdots \wedge dy_n, \qquad \text{on } U_j.$$

On $U_i \cap U_j$, we have

$$\omega = a_j(y_1(x_1, \cdots, x_n), \cdots, y_n(x_1, \cdots, x_n))$$

$$\times J \left(\frac{y_1, \cdots, y_n}{x_1, \cdots, x_n} \right) dx_1 \wedge \cdots \wedge dx_n,$$

where J is the Jacobian of the coordinate transformation. Now,

$$\int_{U_i} \omega F_i = \int F_i a_i \, dx_1 \wedge \cdots \wedge dx_n$$

$$= \int_{B_n} F_i(x_1, \cdots, x_n) a_i(x_1, \cdots, x_n) \, dx_1 \wedge \cdots \wedge dx_n,$$

where we have used the same functional notation for the composition of functions on U_i with the coordinate mapping as for functions on U_i.

We now define

$$\int \omega = \int_M \omega$$

as

$$\int \omega = \sum_{i=1}^m \int_{U_i} \omega F_i.$$

Proposition 3

The above definition of integral is independent of the covering and the partition of unity chosen.

Proof

Let $V_1, \cdots, V_l$ be another covering of M with coordinate neighborhoods and let $G_1, \cdots, G_l$ be the corresponding partition of unity. We must show that

$$\sum_{i=1}^m \int_{U_i} \omega F_i = \sum_{j=1}^l \int_{V_j} \omega G_j.$$

Now,

$$U_1 \cap V_1, \cdots, U_m \cap V_l$$

is a covering of M by coordinate neighborhoods and

$$F_1 G_1, \cdots, F_k G_l$$

is a corresponding partition of unity, for $F_i G_j$ vanishes off

$$U_i \cap V_j, \qquad i = 1, \cdots, m, \quad j = 1, \cdots, l,$$

and

$$\sum_{i=1}^m \sum_{j=1}^l F_i(p) G_j(p) = \sum_{i=1}^m F_i(p) \sum_{j=1}^l G_j(p) = \sum_{i=1}^m F_i(p) = 1,$$

for every $p \in M$. Finally,

$$\sum_{i=1}^m \int_{U_i} \omega F_i = \sum_{i=1}^m \sum_{j=1}^l \int_{U_i \cap V_j} \omega F_i G_j = \sum_{j=1}^l \sum_{i=1}^m \int_{U_i \cap V_j} \omega G_j F_i$$

$$= \sum_{j=1}^l \int_{V_j} \omega G_j \sum_{i=1}^m F_i = \sum_{j=1}^l \int_V \omega G_j. \quad \blacksquare$$

We turn now to the theorem of Stokes. Let M be a compact oriented differentiable n-dimensional manifold. Let $D \subset M$ be a connected open set in M, and let ∂D be the boundary of D. Then ∂D consists of those points $p \in M$ such that every neighborhood of p meets both D and the complement of D. We shall suppose ∂D is a differentiable manifold of dimension $n - 1$. In the proof of the theorem of Stokes we need only assume the manifolds and the forms ω to be of class C'. The orientation in D induces an orientation in ∂D. We shall discuss this briefly later. The theorem of Stokes asserts that for differentiable $(n - 1)$-forms ω on M, we have

$$\int_D d\omega = \int_{\partial D} \omega.$$

We shall prove this theorem under certain conditions on D. We shall first prove that Stokes' theorem is, in fact, a local theorem; i.e., we have

Lemma 3

If for every $p \in \partial D$ there is a coordinate neighborhood U of p such that

$$\phi(U \cap \partial D) = B_{n-1} = [x: x \in B_n, x_n = 0],$$

and

$$\int_D d\omega = \int_{\partial D} \omega$$

holds for every ω whose support is in U (i.e., which vanishes outside a closed subset of U), then

$$\int_D d\omega = \int_{\partial D} \omega,$$

for every ω.

Proof

M has a covering by means of coordinate neighborhoods $U_1, \cdots, U_k$, each of which satisfies the hypothesis of the lemma, or is such that its closure does not meet ∂D. Let $F_1, \cdots, F_k$ be a corresponding partition of unity. Then

$$\int_D d\omega = \sum_{i=1}^{k} \int_D F_i \, d\omega$$

and

$$\int_{\partial D} \omega = \sum_{i=1}^{k} \int_{\partial D} F_i \omega,$$

since the property of the chosen U_i is such that the partition of unity for M induces a partition of unity for ∂D.

The crucial point is that

$$\int_D d\omega = \sum_{i=1}^{k} \int_D d(F_i \omega).$$

This is true since

$$\sum_{i=1}^{k} d(F_i\omega) = \sum_{i=1}^{k} (dF_i \wedge \omega + F_i \, d\omega) = \sum_{i=1}^{k} dF_i \wedge \omega + \sum_{i=1}^{k} F_i \, d\omega$$

$$= \left(\sum_{i=1}^{k} dF_i\right) \wedge \omega + \sum_{i=1}^{k} F_i \, d\omega = \left(d\sum_{i=1}^{k} F_i\right) \wedge \omega + d\omega.$$

But

$$\sum_{i=1}^{k} F_i = 1$$

so that

$$d\sum_{i=1}^{k} F_i = 0.$$

Hence

$$d\omega = \sum_{i=1}^{k} d(F_i\omega).$$

We then have

$$\int_D d\omega = \sum_{i=1}^{k} \int_D F_i \, d\omega = \sum_{i=1}^{k} \int_D d(F_i\omega)$$

$$= \sum_{i=1}^{k} \int_{\partial D} F_i\omega = \int_{\partial D} \omega,$$

since

$$\int_D d(F_i\omega) = \int_{\partial D} F_i\omega,$$

by the hypothesis of the lemma, for those of the $U_1, \cdots, U_k$ which meet ∂D, and

$$\int_D d(F_i\omega) = \int_{\partial D} F_i\omega = 0,$$

for those of the $U_1, \cdots, U_k$ whose closure does not meet ∂D.

Remark   The fact that the U_i are such that

$$\phi_i(U_i \cap \partial D) = B_{n-1}$$

reduces the matter of what orientation is induced in ∂D by the orientation of M to that of what orientation is induced in B_{n-1} by that of B_n, or equivalently by that of the half of B_n, bounded by B_{n-1}, which meets $\phi(D)$.

For the latter case, we may just as well consider the orientation induced in a face of a simplex by the orientation of the simplex itself. An n-simplex S_n, defined by $n + 1$ vertices $(a_0, a_1, \cdots, a_n)$, has two orientations. A rearrangement of the vertices gives the simplex the same orientation if it is obtained from the given one by an even permutation; otherwise it gives the simplex the opposite orientation. The face S_{n-1}, whose vertices are $a_1, \cdots, a_n$, is given the orientation $(a_1, \cdots, a_n)$ by the orientation $(a_0, a_1, \cdots, a_n)$ in S_n.

We now prove

Theorem 4 (Stokes)

If M is a compact, orientable, n-dimensional differentiable manifold, if D is a connected open set in M, with boundary ∂D, and if for each $p \in \partial D$ there is a coordinate neighborhood U of p, with coordinate mapping

$$\phi: U \to B_n$$

such that

$$\phi(U \cap \partial D) = B_{n-1},$$

then, for every differentiable $(n-1)$-form ω on M, we have

$$\int_D d\omega = \int_{\partial D} \omega,$$

where the integrals are over D and ∂D, with the orientations induced by the orientation of M.

Remark Every D whose boundary ∂D is itself a differentiable $(n-1)$-dimensional manifold has the above property, but the proof is beyond the scope of this book.

Proof

If $p \in \partial D$, take U as allowed by the hypothesis of the theorem and suppose the support of ω is in U. If ω and $d\omega$ are expressed in terms of the coordinates of U, for example, if ω is written as

$$\omega = \sum_{i=1}^{n} (-1)^{i-1} a_i \, dx_1 \wedge \cdots \wedge dx_{i-1} \wedge dx_{i+1} \wedge \cdots \wedge dx_n,$$

and if the half of B_n which contains $\phi(D \cap U)$ is extended to an n-cube, one of whose faces is $x_n = 0$, then if $\omega \circ \phi^{-1}$ is extended to be zero, where it is not already defined, the proof that

$$\int_D d\omega = \int_{\partial D} \omega = 0$$

follows by a simple application of iterated integration. ∎

We leave it to the reader to extend Stokes' theorem to the case where M is σ-compact and $D \cup \partial D$ is compact.

11. EXAMPLES

We give some examples of Stokes' theorem. In each of the examples, M will be a euclidean space, and the conditions of Theorem 4 will always be assumed to hold.

Example A Let M be the real line and let D be an open interval $D = (a, b)$, with the designated orientation. The boundary of D will then be $\partial D = \{b\} - \{a\}$. Let ω be a 0-form on M, i.e., a differentiable function f. Its differential $d\omega$ has the representation $f'(x)\, dx$. Thus Stokes' theorem says

$$\int_{(a,b)} f'(x)\, dx = \int_{\partial(a,b)} f = f(b) - f(a).$$

Thus the fundamental theorem of the calculus is a special case of Stokes' theorem.

Example B Let M be euclidean 2-space and let D be a bounded connected open set with boundary ∂D. Let $\omega = p\, dx + q\, dy$ be a 1-form in M. Then

$$d\omega = \left(\frac{\partial q}{\partial x} - \frac{\partial p}{\partial y} \right) dx \wedge dy.$$

Then Stokes' theorem asserts

$$\iint_{D} \left(\frac{\partial q}{\partial x} - \frac{\partial p}{\partial y} \right) dx \wedge dy = \int_{\partial D} p\, dx + q\, dy.$$

This is known as Green's theorem.

Example C Let M be euclidean 3-space and let D be a bounded connected open set with boundary ∂D. Let

$$\omega = p\, dx \wedge dy + q\, dy \wedge dz + r\, dz \wedge dx.$$

Then

$$d\omega = \left(\frac{\partial p}{\partial z} + \frac{\partial q}{\partial x} + \frac{\partial r}{\partial y} \right) dx \wedge dy \wedge dz.$$

Now, Stokes' theorem says

$$\iiint_{D} \left(\frac{\partial p}{\partial z} + \frac{\partial q}{\partial x} + \frac{\partial r}{\partial y} \right) dx \wedge dy \wedge dz$$

$$= \iint_{\partial D} p\, dx \wedge dy + q\, dy \wedge dz + r\, dz \wedge dx.$$

This is known as the divergence theorem.

12. PERIODS OF 1-FORMS

We now give a brief discussion of the converse of Theorem 3. We consider only 1-forms on compact, oriented, two-dimensional manifolds.

A 1-form is **locally exact** if it is exact in every coordinate neighborhood U. Thus, if (x_1, x_2) are the coordinates in U and if in these coordinates

$$\omega = a \, dx_1 + b \, dx_2,$$

there is a function f (a 0-form) such that

$$df = \omega$$

in U. A necessary and sufficient condition for this is

$$\frac{\partial a}{\partial x_2} = \frac{\partial b}{\partial x_1}.$$

The reader may easily verify that ω is locally exact if and only if $d\omega = 0$, i.e., ω is closed.

In order to exhibit closed 1-forms which are not exact, it is convenient to introduce the notion of **period of a differential form** ω on a closed curve C in the manifold M. We shall consider just simple closed curves, i.e., homeomorphic images of a circle. The period is the number

$$\int_C \omega.$$

If ω is exact, then all the periods are 0, since $\omega = df$ implies

$$\int_C \omega = \int_C df = f(p) - f(p) = 0,$$

for any closed C and $p \in C$.

The converse also holds. Suppose

$$\int_C \omega = 0,$$

for every closed C. Then for fixed $p \in M$,

$$\int_p^q \omega = f(q)$$

is independent of the path joining p to q, so that f is defined. It remains only to show that $df = \omega$. This is done by showing that for any $q \in M$ and local coordinates (x_1, x_2) around q, with

$$\omega = a \, dx_1 + b \, dx_2,$$

we have

$$\frac{\partial f}{\partial x_1} = a, \quad \frac{\partial f}{\partial x_2} = b.$$

The details are left to the reader.

It remains for us to give an example of a closed 1-form which is not exact. For this purpose, we consider the torus. Coordinates on the torus are given by angles (θ, ϕ). However, these coordinates are not single-valued. Thus, we must cover the torus by a finite set of coordinate neighborhoods $U_1, \cdots, U_k$, with coordinates (θ_i, ϕ_i) for U_i, $i = 1, \cdots, k$, such that on each $U_i \cap U_j$, either

$$\theta_i = \theta_j, \theta_i = \theta_j + 2\pi \quad \text{or} \quad \theta_i = \theta_j - 2\pi,$$

and similarly for $\dot{\phi}_i$ and ϕ_j.

Thus the torus is a differentiable manifold (of class C^∞) with these coordinate neighborhoods and mappings. Then

$$d\theta \quad \text{and} \quad d\phi$$

are 1-forms on the torus. There is a closed C (see Fig. 7), for which

$$\int_C d\theta = 2\pi.$$

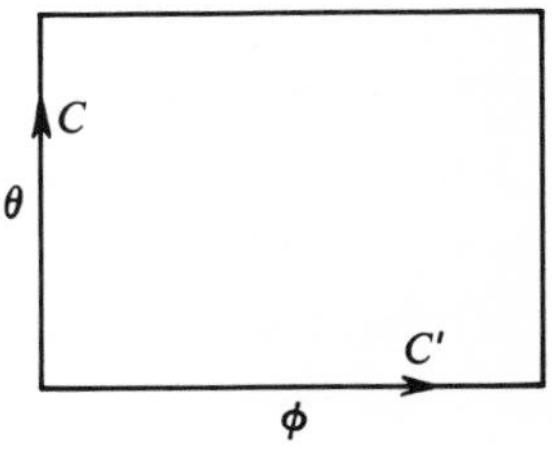

But $d\theta$ is evidently locally exact, hence closed. There is another closed curve C' for which

$$\int_{C'} d\varphi = 2\pi.$$

Figure 7

It can be shown that for any closed 1-form ω there are constants a, b such that

$$\omega - (a\, d\theta + b\, d\phi)$$

is exact.

EXERCISES

1.1 Show that a line interval and a circle are not homeomorphic.

1.2 Show that euclidean 1-space and euclidean 2-space are not homeomorphic.

1.3 Show that the closed unit disk is not homeomorphic to the closed unit 2-sphere.

1.4 Show that the torus is not homeomorphic to the 2-sphere.

2.1 Prove that a closed interval is not a manifold.

2.2 Prove that for a torus the collection $\mathscr{U}$ can be taken to have 3 members, but it cannot be taken to have 2 members.

2.3 Answer the same sort of question for the projective plane.

3.1 Define a covering for the 2-sphere for which it is a differentiable manifold. Prove.

3.2 Define a covering for the torus for which it is a differentiable manifold. Prove.

3.3 Show that the projective plane is not orientable.

3.4 Discuss the k-holed torus. Obtain a representation in the plane and make all proper identifications.

3.5 Show that the k-holed torus is a differentiable manifold.

3.6 Define the product $M \times N$ of differentiable manifolds M and N.

3.7 Discuss the differentiable functions on $M \times N$ in terms of those on M and on N.

4.1 Show that in a differentiable manifold there is a maximal set of coordinate neighborhoods.

4.2 Show that every compact differentiable manifold M can be imbedded in a euclidean space (whose dimension may be considerably higher than that of the manifold).

Outline of solution: Let $U_1, \cdots, U_k$ and $V_1, \cdots, V_k$ be coverings of M by coordinate neighborhoods, with $V_i \subset U_i$, $i = 1, \cdots, k$. Let f_i be 1 on V_i, 0 on $M \sim U_i$, and between 0 and 1 elsewhere. For each $i = 1, \ldots, k$, consider the $(n + 1)$-functions $x_{i0}(p) = f_i(p)$,

$$x_{ij}(p) = \begin{cases} \phi_{ij}(p)f_i(p), & p \in U_i, \\ 0, & p \notin U_i, \end{cases}$$

where $\phi_i = (\phi_{i1}, \cdots, \phi_{in})$ is the coordinate mapping function. This mapping gives the desired imbedding in

$$R^{(n+1)k}.$$

5.1 Given points p, q in the open n-ball B_n, prove that there is a diffeomorphism $f: B_n \to B_n$ with $f(p) = q$.

5.2 Given points

$$p_1, \cdots, p_k; q_1, \cdots, q_k$$

in B_2, show that there is a diffeomorphism

$$f: B_2 \to B_2$$

with

$$f(p_i) = q_i, \quad i = 1, \cdots, k.$$

5.3 Given the arc
$$x^2 + (y - \tfrac{1}{2})^2 = 1$$
in the ball B_2, show that there is a diffeomorphism $f: B_2 \to B_2$ which takes this arc onto the segment $y = 0$.

5.4 Show that the function $f: R^n \to R$, described in the text, has the properties stated there.

6.1 A vector field is an association of an element of $\mathscr{T}_p$ with every $p \in M$. Define differentiability of a vector field in terms of its action on elements of $\mathscr{F}_p$.

6.2 Show that, with operations defined in the appropriate way, the vector fields on M form a vector space.

6.3 Find 2 linearly independent vector fields on the torus.

6.4 For vector fields X, Y define the braces by
$$[X, Y]f = X(Yf) - Y(Xf), f \in \mathscr{F}_p.$$
Show that $[X, Y]$ is a vector field satisfying

(a) $\qquad [X, Y] = -[Y, X],$

(b) $\qquad [[X, Y], Z] + [[Y, Z], X] + [[Z, X], Y] = 0.$

6.5 Show that, for every differentiable mapping ϕ of one manifold into another, there is an induced mapping ϕ^* of tangent spaces, such that, for each $p \in M$,
$$\phi^*: \mathscr{T}_{\phi(p)} \to \mathscr{T}_p.$$

8.1 Show that vectors
$$x^1, \cdots, x^n \in V$$
are linearly independent if and only if
$$x^1 \wedge \cdots \wedge x^n \neq 0.$$

8.2 Give a detailed proof that $G(V)$, as constructed in the text, is an algebra.

8.3 Prove Proposition 1 for the case $p = 3$, $n = 5$.

8.4 Prove Proposition 1 when $p = n$.

9.1 Show that a mapping $\phi: M \to N$ induces a mapping ϕ_* between differential forms.

9.2 Show that ϕ_* commutes with d.

9.3 Find a differential $(n - 1)$-form ω in R^n such that
$$d\omega = dx_1 \wedge \cdots \wedge dx_n.$$

10.1 Show that for any $(n - 1)$-form ω on the n-sphere S,
$$\int_S d\omega = 0.$$

10.2 Show that, with the coordinate neighborhoods of the lemma, a partition of unity in M does, indeed, induce a partition of unity in ∂D.

10.3 Give the details of the proof of Theorem 4, not given in the text, taking into account the induced orientation of ∂D.

10.4 Extend the proof of Stokes' theorem to the case of a σ-compact manifold, where $D \cup \partial D$ is compact.

10.5 Prove Stokes' theorem for differential forms in the plane where ∂D is composed of finitely many continuously differentiable arcs.

10.6 Find out what is meant by a standard simplex. Prove Stokes' theorem for the case where D is a standard simplex.

10.7 Find out what is meant by a singular chain and, using exercise 9.2, prove Stokes' theorem for this case.

10.8 Let f be a continuously differentiable function on $(-1, 1)$ such that $|f(x)| < 1$ for all $x \in (-1, 1)$, and let $[f]$ be that part of the graph of f which lies in B_2. Show that there is a diffeomorphism of B_2 onto itself which takes $[f]$ onto the line $x_2 = 0$.

11.1 In Example 2, show that Stokes' theorem holds if D is convex by approximating the boundary of D in a manner suitable for applying the theorem we already have.

12.1 Verify that a 1-form on a 2-manifold is locally exact if and only if it is closed.

12.2 Complete the proof that if all the periods of a 1-form on a manifold are zero then the form is exact.

12.3 Show that every closed 1-form on a 2-sphere is exact.

12.4 In the example of the torus, exhibit coordinate neighborhoods $U_1, \cdots, U_k$ as required.

12.5 Given any real numbers a and b, show that there is a closed form whose periods on C_1 and C_2 are a and b, respectively (C_1 and C_2 are the C and C' of Fig. 7).

12.6 Show that if a closed form has periods zero on C_1 and C_2, then its periods are zero on every closed curve C, so that it is exact.

12.7 Show that for any closed 1-form ω on the torus, there are constants a and b such that

$$\omega - (a \, d\theta + b \, d\phi)$$

is exact.

CHAPTER VI

VECTOR ANALYSIS

There is a subject, called vector analysis, which is largely concerned with special forms of Stokes' theorem. It is the purpose of this very brief chapter to give a glimpse into this topic and the related subject of harmonic functions.

1. CROSS PRODUCT

We shall be concerned, in the first sections of this chapter, with euclidean 3-space R^3. We suppose R^3 has an orthonormal basis which we now designate by i, j, k. The corresponding coordinates are designated by (x, y, z). An arbitrary vector will then have the form

$$\xi = xi + yj + zk.$$

The scalar product of vectors

$$\xi_1 = x_1 i + y_1 j + z_1 k$$

and

$$\xi_2 = x_2 i + y_2 j + z_2 k$$

is given by

$$\xi_1 \cdot \xi_2 = x_1 x_2 + y_1 y_2 + z_1 z_2.$$

Note that this is the scalar product introduced in Chapter I, but with a different notation.

In this special topic, it is also essential to have a **vector product.** This is defined by

$$\xi_1 \times \xi_2 = \begin{vmatrix} i & j & k \\ x_1 & y_1 & z_1 \\ x_2 & y_2 & z_2 \end{vmatrix}$$

$$= (y_1 z_2 - z_1 y_2)i + (z_1 x_2 - x_1 z_2)j + (x_1 y_2 - y_1 x_2)k.$$

Vectors will also be considered in euclidean 2-space.

Proposition 1

$\xi_1 \times \xi_2$ *is orthogonal to* ξ_1 *and* ξ_2.

Proof

For ξ_1, we have

$$(\xi_1 \times \xi_2) \cdot \xi_1 = \begin{vmatrix} i & j & k \\ x_1 & y_1 & z_1 \\ x_2 & y_2 & z_2 \end{vmatrix} \cdot (x_1 i + y_1 j + z_1 k) = \begin{vmatrix} x_1 & y_1 & z_1 \\ x_1 & y_1 & z_1 \\ x_2 & y_2 & z_2 \end{vmatrix} = 0.$$

Similarly,

$$(\xi_1 \times \xi_2) \cdot \xi_2 = 0. \quad\blacksquare$$

In particular, it is of interest to consider the special cross products

$$i \times j, \quad j \times k, \quad \text{and} \quad k \times i.$$

$$i \times j = \begin{vmatrix} i & j & k \\ 1 & 0 & 0 \\ 0 & 1 & 0 \end{vmatrix} = k.$$

In similar fashion,

$$j \times k = i \quad \text{and} \quad k \times i = j.$$

We also have

Proposition 2

For any two vectors ξ_1 and ξ_2,

$$\xi_1 \times \xi_2 = -\xi_2 \times \xi_1.$$

We leave the proof to the reader, as well as the proof that

$$\xi_1 \times \xi_2 = 0,$$

the zero vector, if and only if

$$\xi_1 = a\xi_2 \quad \text{or} \quad \xi_2 = a\xi_1,$$

a real.

Finally, we leave it to the reader to find an expression for the magnitude of the vector $\xi_1 \times \xi_2$.

2. GRADIENT, DIVERGENCE, CURL

Let $G \subset R^3$ be an open set and let

$$f \colon G \to R$$

be a differentiable function. Such a function is called a **scalar field.** The triple

$$\left(\frac{\partial f}{\partial x}, \frac{\partial f}{\partial y}, \frac{\partial f}{\partial z} \right)$$

is known as the **gradient** of f. In vector notation, the gradient is written

$$\operatorname{grad} f = \frac{\partial f}{\partial x} i + \frac{\partial f}{\partial y} j + \frac{\partial f}{\partial z} k.$$

It is a vector valued function, or **vector field,** on G. The operator

$$\nabla = \frac{\partial}{\partial x} i + \frac{\partial}{\partial y} j + \frac{\partial}{\partial z} k$$

is useful in this connection. Then

$$\nabla f = \frac{\partial f}{\partial x} i + \frac{\partial f}{\partial y} j + \frac{\partial}{\partial x} k.$$

The gradient or ∇ operator defines a mapping which takes scalar functions into vector functions. We consider two further operators, one of which takes vector functions into scalar functions, the other of which takes vector functions into vector functions.

Let f be a vector function on G. Then

$$f = f_1 i + f_2 j + f_3 k.$$

The **divergence** of f is the scalar function

$$\operatorname{div} f = \frac{\partial f_1}{\partial x} + \frac{\partial f_2}{\partial y} + \frac{\partial f_3}{\partial z}.$$

In terms of ∇, it may be written as

$$\nabla \cdot f = \operatorname{div} f = \left(\frac{\partial}{\partial x} i + \frac{\partial}{\partial y} j + \frac{\partial}{\partial z} k \right) \cdot (f_1 i + f_2 j + f_3 k)$$

$$= \frac{\partial f_1}{\partial x} + \frac{\partial f_2}{\partial y} + \frac{\partial f_3}{\partial z}.$$

The vector function associated with $f = f_1 i + f_2 j + f_3 k$ is called the **curl** of f. It is defined as

$$\operatorname{curl} f = \begin{vmatrix} i & j & k \\ \dfrac{\partial}{\partial x} & \dfrac{\partial}{\partial y} & \dfrac{\partial}{\partial z} \\ f_1 & f_2 & f_3 \end{vmatrix}$$

$$= \left(\frac{\partial f_3}{\partial y} - \frac{\partial f_2}{\partial z} \right) i + \left(\frac{\partial f_1}{\partial z} - \frac{\partial f_3}{\partial x} \right) j + \left(\frac{\partial f_2}{\partial x} - \frac{\partial f_1}{\partial y} \right) k.$$

Clearly, the symbol $\nabla \times f$ represents the curl of f. The symbol ∇ is thus featured in the three important operators of vector analysis.

∇f is the gradient of a scalar function f.

$\nabla \cdot f$ is the divergence of a vector function f.

$\nabla \times f$ is the curl of a vector function f.

Another important operator, taking scalar functions into scalar functions, is the **Laplacian.** This acts on twice-differentiable functions and takes each such f into

$$\frac{\partial^2 f}{\partial x^2} + \frac{\partial^2 f}{\partial y^2} + \frac{\partial^2 f}{\partial z^2}.$$

The Laplacian is representable as

$$\nabla^2 f = \nabla \cdot \nabla f = \frac{\partial^2 f}{\partial x^2} + \frac{\partial^2 f}{\partial y^2} + \frac{\partial^2 f}{\partial z^2}.$$

It is often written as Δf. Laplace's equation

$$\Delta f = 0$$

arises in many places in pure and applied mathematics.

3. FORMS OF STOKES' THEOREM

We now shall express line and surface integrals in vector notation.

For line integrals, the integrand is a 1-form which may be expressed, in terms of the (x, y, z)-coordinates, as

$$\omega = f_1\, dx + f_2\, dy + dz.$$

If ω is to be integrated along a differentiable curve C, if $t(p)$ is the unit tangent vector to C at $p \in C$, if

$$f = f_1 i + f_2 j + f_3 k$$

and

$$r = xi + yj + zk,$$

then

$$\omega = f \cdot dr = f \cdot t \, ds,$$

where $|dr| = ds$. It follows that we have

$$\int_C \omega = \int_C f \cdot dr = \int_C f \cdot t \, ds.$$

For surface integrals, consider a surface which is a differentiable mapping of a portion of the (u, v)-plane into euclidean 3-space, given by functions

$$x = x(u, v),$$
$$y = y(u, v),$$
$$z = z(u, v).$$

We consider the 2-form

$$\omega = \left(\frac{\partial f_3}{\partial y} - \frac{\partial f_2}{\partial z}\right) dy \wedge dz + \left(\frac{\partial f_1}{\partial z} - \frac{\partial f_3}{\partial x}\right) dz \wedge dx + \left(\frac{\partial f_2}{\partial x} - \frac{\partial f_1}{\partial y}\right) dx \wedge dy$$

$$= \text{curl} f \cdot (dy \wedge dz \, i + dz \wedge dxj + dx \wedge dyk).$$

We obtain an expression for the vector on the right. Choose a point on the given surface, corresponding to a fixed (u, v). The direction of the normal to the surface at this point is

$$\frac{\partial(y, z)}{\partial(u, v)} : \frac{\partial(z, x)}{\partial(u, v)} : \frac{\partial(x, y)}{\partial(u, v)},$$

where, for example,

$$\frac{\partial(y, z)}{\partial(u, v)}$$

is the Jacobian of the mapping

$$y = y(u, v), \quad z = z(u, v).$$

We then obtain

$$\omega = \operatorname{curl} f \cdot \left[\frac{\partial(y, z)}{\partial(u, v)} i + \frac{\partial(z, x)}{\partial(u, v)} j + \frac{\partial(x, y)}{\partial(u, v)} k \right] du \wedge dv$$

$$= \operatorname{curl} f \cdot \frac{\dfrac{\partial(y, z)}{\partial(u, v)} i + \dfrac{\partial(z, x)}{\partial(u, v)} j + \dfrac{\partial(x, y)}{\partial(u, v)} k}{\left[\left\{ \dfrac{\partial(y, z)}{\partial(u, v)} \right\}^2 + \left\{ \dfrac{\partial(z, x)}{\partial(u, v)} \right\}^2 + \left\{ \dfrac{\partial(x, y)}{\partial(u, v)} \right\}^2 \right]^{1/2}} dA,$$

by the integral formula for the area of a surface. Since the complicated quotient is merely the unit normal vector n to the surface, we obtain

$$\omega = \operatorname{curl} f \cdot n \, dA.$$

Accordingly, surface integrals take the form

$$\iint_S \omega = \iint \operatorname{curl} f \cdot n \, dA.$$

We are now ready to write some integral formulas in vector notation.

(a) Let M be a two-dimensional differentiable manifold imbedded in 3-space, let D be a connected open set in M, whose boundary is a differentiable manifold, and let ω be a 1-form on M, all of class C'. Then

$$\int_D d\omega = \int_{\partial D} \omega.$$

But in accordance with the above discussion,

$$d\omega = \operatorname{curl} f \cdot n \, dA$$

and

$$\omega = f \cdot dr.$$

Thus Stokes' theorem takes the form

$$\int_D \operatorname{curl} f \cdot n \, dA = \int_{\partial D} f \cdot dr.$$

(b) As a second example, we obtain the divergence theorem. Let D be an open connected set in euclidean 3-space, whose boundary ∂D is a differentiable manifold, and let ω be a differentiable 2-form on R^3. In vector notation, the divergence theorem becomes

$$\iiint_D \nabla \cdot f \, dV = \iint_{\partial D} f \cdot n \, dA.$$

4. DEFINITION OF HARMONIC FUNCTION

We recall that Green's theorem asserts that

$$\iint_D \left(\frac{\partial q}{\partial x} - \frac{\partial p}{\partial y} \right) dx \wedge dy = \int_{\partial D} p \, dx + q \, dy,$$

where D is a bounded connected open set, ∂D is a C'-manifold of dimension one, and p and q are of class C'.

By assuming u and v of class C^2 and making proper choices of p and q, we obtain

$$\iint_D \left((u \, \Delta v + \operatorname{grad} u \cdot \operatorname{grad} v) \, dx \, dy = \int_{\partial D} u \frac{\partial v}{\partial n} \, ds,$$

where s is arc length, $\partial v/\partial n$ is the directional derivative of v in the direction of the outward normal to D, and

$$\Delta v = \frac{\partial^2 v}{\partial x^2} + \frac{\partial^2 v}{\partial y^2}.$$

By interchanging the roles of u and v we obtain the important

$$(*) \qquad \iint_D (u \, \Delta v - v \, \Delta u) \, dx \, dy = \int_{\partial D} \left(u \frac{\partial v}{\partial n} - v \frac{\partial u}{\partial n} \right) ds.$$

A function u is said to be **harmonic** on an open set G if

$$\Delta u = 0, \qquad \text{on } G.$$

If u and v are harmonic, $(*)$ becomes

$$\int_{\partial D} \left(u \frac{\partial v}{\partial n} - v \frac{\partial u}{\partial n} \right) ds = 0.$$

If in particular, along with u being harmonic, $v = 1$, we obtain

$$\int_{\partial D} \frac{\partial u}{\partial n} \, ds = 0.$$

5. MEAN VALUE AND MAXIMUM PRINCIPLE

The harmonic functions constitute one of the most remarkable classes of functions in mathematics, and we close by giving a small sampling of their properties.

Suppose u is harmonic in an open set G, and D is a disk with center (x_0, y_0) and radius r such that $D \cup \partial D \subset G$. We prove

Theorem 1

Under the conditions given above

$$u(x_0, y_0) = \frac{1}{2\pi} \int_0^{2\pi} u(x_0 + r \cos \theta, y_0 + r \sin \theta) \, d\theta.$$

Proof

If v is the constant function 1, for $0 < \rho < r$, we obtain for the disk D_ρ of radius ρ and center (x_0, y_0),

$$0 = \int_{\partial D_\rho} \frac{\partial u}{\partial n} \, ds = \int_0^{2\pi} \frac{\partial u}{\partial \rho} (x_0 + \rho \cos \theta, y_0 + \rho \sin \theta) \, d\theta$$

$$= \frac{\partial}{\partial \rho} \int_0^{2\pi} u(x_0 + \rho \cos \theta, y_0 + \rho \sin \theta) \, d\theta.$$

It follows that

$$\int_0^{2\pi} u(x_0 + \rho \cos \theta, y_0 + \rho \sin \theta) \, d\theta = g(\rho)$$

is constant on $0 < \rho < r$. Since this integral is continuous in ρ, we obtain

$$g(\rho) = \int_0^{2\pi} u(x_0 + r \cos \theta, y_0 + r \sin \theta) \, d\theta.$$

On the other hand, the continuity of u at (x_0, y_0) implies that

$$u(x_0, y_0) = \lim_{\rho \to 0} \frac{1}{2\pi} \int_0^{2\pi} u(x_0 + \rho \cos \theta, y_0 + \rho \sin \theta) \, d\theta. \quad \blacksquare$$

This proves the theorem.

We may state

Corollary 1

If u is harmonic in an open set G and if $D \subset G$ is a disk with center (x_0, y_0) and radius r, then

$$u(x_0, y_0) = \frac{1}{\pi r^2} \iint_D u(x, y) \, dx \, dy.$$

We leave the proof to the reader. Theorem 1 and its corollary express the fact that harmonic functions have a mean-value property. We shall show later that a converse also holds.

In the meantime, another easily derived property of harmonic functions is the maximum principle, which we state in the following form.

Theorem 2

If u is harmonic in a connected open set G, and is not constant, then for every $(x, y) \in G$, there is an $(x', y') \in G$ such that

$$u(x', y') > u(x, y).$$

Proof

Suppose $(x, y) \in G$, and $u(x, y) \geq u(x', y')$ for every $(x', y') \in G$. Fix $(x', y') \in G$. By Theorem 17, Chapter I, there is a finite chain of disks $D_1, D_2, \cdots, D_k$ in G such that the center of D_i is in D_{i-1}, $i = 2, \cdots, k$, (x, y) is the center of D_1, and (x', y') is the center of D_k.

Since $u(\xi, \eta) \leq u(x, y)$ for every $(\xi, \eta) \in D_1$, Corollary 1 implies $u(\xi, \eta) = u(x, y)$ for every $(\xi, \eta) \in D_1$. Let (x_2, y_2) be the center of D_2. By the same argument as for D_1, $u(\xi, \eta) = u(x_2, y_2) = u(x, y)$ for every $(\xi, \eta) \in D_2$. By induction, we obtain $u(x', y') = u(x, y)$. ∎

6. POISSON INTEGRAL FORMULA

In this section, we show that every continuous function on the unit circle has a continuous extension to a function which is harmonic on the open unit disk. Moreover, this extension is unique.

Thus, let f be a continuous function on the real line, of period 2π. We wish to investigate the existence and uniqueness of a continuous function u on the unit disk such that

$$u(r \cos \theta, r \sin \theta)$$

is harmonic for $0 \leq r < 1$ and

$$u(\cos \theta, \sin \theta) = f(\theta).$$

Suppose u and v both satisfy the conditions. Then $w = u - v$ is harmonic for $0 \leq r < 1$, and

$$w(\cos \theta, \sin \theta) = 0, \qquad 0 \leq \theta < 2\pi.$$

Theorem 2 implies $w \leq 0$. But clearly a minimum principle also is valid, so that $w \geq 0$. Hence

$$u = v,$$

and the uniqueness is established.

The existence is much more difficult. We consider the Fourier series

$$\frac{a_0}{2} + \sum_{n=1}^{\infty} (a_n \cos n\theta + b_n \sin n\theta)$$

of the continuous periodic function f.

We consider a function suggested by this series, namely, for $0 \leq r < 1$,

$$u(r \cos \theta, r \sin \theta) = \frac{a_0}{2} + \sum_{n=1}^{\infty} (a_n \cos n\theta + b_n \sin n\theta) r^n$$

$$= \frac{1}{\pi} \int_0^{2\pi} f(\phi) \left[\frac{1}{2} + \sum_{n=1}^{\infty} r^n \cos n\, (\theta - \phi) \right] d\phi$$

$$= \frac{1}{2\pi} \int_0^{2\pi} f(\phi)\ Re \left[\frac{1 + re^{i(\theta - \phi)}}{1 - re^{i(\theta - \phi)}} \right] d\phi$$

$$= \frac{1}{2\pi} \int_0^{2\pi} f(\phi)\ \frac{1 - r^2}{1 - 2r \cos(\theta - \phi) + r^2}\, d\phi.$$

This merely suggests the above function u as a possibility. We now show that it actually furnishes the solution to our problem.

We must first show that u is harmonic in the open unit disk. This is a direct computation, involving the second partial derivatives, and is left to the reader.

Finally, we must show that for every θ,

$$\lim_{\substack{\phi \to \theta \\ r \to 1}} u(r \cos \phi, r \sin \phi) = f(\theta).$$

It is enough, by symmetry, to prove this for $\theta = 0$.

We first note that

$$\frac{1}{2\pi} \int_0^{2\pi} \frac{1 - r^2}{1 - 2r \cos(\theta - \phi) + r^2}\, d\phi = 1,$$

for all ϕ and all $r < 1$. We thus have

$$|u(r \cos \phi, r \sin \phi) - f(0)| \leq \frac{1}{2\pi} \int_0^{2\pi} |f(\theta) - f(0)|$$

$$\times \frac{1 - r^2}{1 - 2r \cos(\theta - \phi) + r^2}\, d\phi.$$

Let $\epsilon > 0$. Since f is continuous, there is a $\delta > 0$ such that $|\theta| < \delta$ implies $|f(\theta) - f(0)| < \epsilon$. We then obtain

$$|u(r \cos \phi, r \sin \phi) - f(0)| \leq \epsilon + \frac{M}{\pi} \int_\delta^{2\pi - \delta} \frac{1 - r^2}{1 - 2r \cos (\theta - \phi) + r^2} \, d\theta,$$

where

$$M = \max \, [|f(\theta)| : \theta \in [0, 2\pi]].$$

We now need only show that the integral on the right converges to zero as ϕ goes to zero and r goes to 1. For $|\phi| \leq \delta/2$,

$$\frac{1}{2\pi} \int_\delta^{2\pi - \delta} \frac{1 - r^2}{1 - 2r \cos (\theta - \phi) + r^2} \, d\theta < \frac{1 - r^2}{(1 - \cos \delta/2)^2},$$

and this evidently goes to zero as r goes to 1.

This completes the proof of

Theorem 3

If f is a continuous real function of period 2π, the function

$$u(r \cos \theta, r \sin \theta) = \frac{1}{2\pi} \int_0^{2\pi} f(\phi) \frac{1 - r^2}{1 - 2r \cos (\theta - \phi) + r^2} \, d\phi$$

is harmonic on the open unit disk and is such that, for every $\theta \in [0, 2\pi]$,

$$\lim_{\substack{\phi \to \theta \\ r \to 1}} u(r \cos \phi, r \sin \phi) = f(\theta).$$

Moreover, it is the only function with this property.

The formula which gives u is called the **Poisson integral formula.**

Corollary 1

If u is harmonic on the disk of radius $\rho > 0$ and if $0 < R < \rho$, then

$$u(r \cos \theta, r \sin \theta) = \frac{1}{2\pi} \int_0^{2\pi} u(R \cos \phi, R \sin \phi)$$

$$\times \frac{R^2 - r^2}{R^2 - 2rR \cos (\theta - \phi) + r^2} \, d\phi,$$

for $0 \leq r < R$.

Proof

For $R = 1$, this is an obvious consequence of the theorem. We leave the minor adjustments needed for the general case to the reader. ∎

Corollary 2

If G is a connected open set and u is continuous on G and has the mean-value property, then u is harmonic.

Proof

Let D be a disk such that $D \cup \partial D \subset G$. Let v be the harmonic function on D which is extended to a continuous function on $D \cup \partial D$ which agrees with u on ∂D. Let $w = u - v$. Then w has the mean-value property in D and is zero on ∂D. But the mean-value property on D implies that either w is constant or assumes its maximum and minimum on ∂D. It follows that w is zero on D, so that $u = v$ and u is harmonic. ∎

7. HARNACK'S CONVERGENCE THEOREM

An important, but easy to prove, consequence of the Poisson integral formula is the inequality

$$\frac{R-r}{R+r} u(x_0, y_0) \leq u(x, y) \leq \frac{R+r}{R-r} u(x_0, y_0),$$

where u is harmonic in an open disk of radius R and center (x_0, y_0), and the distance of (x, y) from (x_0, y_0) is $r < R$. We leave the proof to the reader.

This inequality has a remarkable consequence, known as Harnack's convergence theorem.

Theorem 4

Let G be an open connected set. If $\{u_n\}$ is a sequence of harmonic functions in G, is nondecreasing at every point in G, and is bounded at one point in G, then it is bounded at every point in G, the convergence is uniform on every compact subset of G, and the limit is harmonic in G.

Proof

Suppose the sequence is bounded at $(x_0, y_0) \in G$. The sequence $\{u_n(x_0, y_0)\}$ converges.

Let R be the radius of the largest disk $D \subset G$ with center (x_0, y_0). Let $\epsilon > 0$. There is an N such that $n > m > N$ implies

$$0 \leq u_n(x_0, y_0) - u_m(x_0, y_0) < \epsilon \frac{R-r}{R+r},$$

where $r < R$ is given.

Let (x, y) be such that

$$[(x - x_0)^2 + (y - y_0)^2]^{1/2} = r.$$

Then, by the above inequality, since $u_n - u_m$ is harmonic, we have

$$0 \le u_n(x, y) - u_m(x, y) \le \frac{R + r}{R - r}\frac{R - r}{R + r}\epsilon.$$

It follows that $\{u_n\}$ converges uniformly to a function u on the closed disk of center (x_0, y_0) and radius r.

A straightforward application of Corollary 1 to Theorem 3, and then of Theorem 3 itself, shows that u is harmonic in the open disk of center (x_0, y_0) and radius r.

That $\{u_n\}$ converges to a harmonic function on all of G and that the convergence is uniform on compact subsets of G follow from the fact that r can be taken as near to R as we wish and that each point in G can be reached by a finite chain of such disks. ∎

EXERCISES

1.1 Find an expression for the magnitude of the vector product of two vectors.

1.2 Verify that

(a) $(\xi_1 \times \xi_2) \times (\xi_3 \times \xi_4) = [\xi_1 \cdot (\xi_2 \times \xi_4)]\xi_3 - [\xi_1 \cdot (\xi_2 \times \xi_3)]\xi_4.$

(b) $[(\xi_1 \times \xi_2) \times \xi_3] \cdot \xi_4 = (\xi_1 \cdot \xi_3)(\xi_2 \cdot \xi_4) - (\xi_2 \cdot \xi_3)(\xi_1 \cdot \xi_4).$

2.1 Show that

$$\text{curl } r = 0,$$

where $r = xi + yj + zk$.

2.2 If grad $f = 0$ everywhere in an open set G, show that f is constant in G.

2.3 Verify:

(a) $\nabla \times (\nabla f) = 0.$

(b) $\nabla \times (\nabla \times f) = \nabla(\nabla \cdot f) - \Delta f.$

2.4 Show that the direction of grad f is that in which the directional derivative of f is maximized and that the magnitude of grad f is the maximum directional derivative.

3.1 Show that

$$\frac{\partial}{\partial x}\nabla f = \nabla \frac{\partial f}{\partial x}.$$

3.2 Let G be a convex open set. Show that f is a gradient vector in G if and only if $\nabla \times f = 0$.

3.3 Use Green's theorem to give a formula for the area of a connected open set in terms of an integral around its boundary.

4.1 Prove the formula

$$\iint_D (u\Delta v - v\Delta u)\, dx\, dy = \iint_{\partial D} \left(u\, \frac{\partial v}{\partial n} - v\, \frac{\partial u}{\partial n} \right) ds.$$

6.1 Let $S_n(f)$ be the nth partial sum of the Fourier series of a continuous f of period 2π. Let

$$\sigma_n(f) = \frac{1}{n+1}\, (S_0(f) + \cdots + S_n(f)).$$

Show that $\sigma_n(f)$ converges uniformly to f.

6.2 Make the changes in Theorem 3 which are necessary to obtain Corollary 1.

6.3 If f and $g \circ f$ are both harmonic in a connected open set and if f is not constant, show that g is linear.

6.4 Show that a harmonic function is infinitely differentiable.

7.1 For harmonic functions of one variable, present that portion of the theory given in the text for harmonic functions of two variables.

7.2 Prove the inequality

$$\frac{R-r}{R+r}\, u(x_0, y_0) \le u(x, y) \le \frac{R+r}{R-r}\, u(x_0, y_0).$$

7.3 Show that if a harmonic function on the whole plane is bounded, then it is constant.

7.4 A function u, on a connected open set G, is said to be subharmonic if it is continuous and if for every $(x_0, y_0) \in G$ and r small enough,

$$u(x_0, y_0) \le \frac{1}{2\pi} \int_0^{2\pi} u(x_0 + r\cos\theta,\, y_0 + r\sin\theta)\, d\theta.$$

7.5 State and prove a maximum principle for subharmonic functions.

7.6 Show that the square of a harmonic function need not be harmonic.

7.7 Which functions are harmonic along with their squares?

7.8 Show that the square of a nonnegative subharmonic function is subharmonic.

REFERENCES

G. Birkhoff and S. MacLane, *A Survey of Modern Algebra*, (rev. ed.), New York, Macmillan, 1953.

R. C. Buck, *Advanced Calculus*, New York, McGraw-Hill, 1956.

R. H. Crowell and R. E. Williamson, *Calculus of Vector Functions*, Englewood Cliffs, N.J., Prentice-Hall, 1962.

H. Flanders, *Differential Forms*, New York, Academic Press, 1963.

W. Fleming, *Functions of Several Variables*, Cambridge, Addison-Wesley, 1965.

J. R. Munkres, *Elementary Differential Topology*, Princeton, N.J., Princeton University Press, 1961.

H. K. Nickerson, D. C. Spencer, and N. E. Steenrod, *Advanced Calculus*, Princeton, N.J., van Nostrand, 1959.

S. Sternberg, *Lectures in Differential Geometry*, Englewood Cliffs, N.J., Prentice-Hall, 1964.

H. Whitney, *Geometric Integration Theory*, Princeton, N.J., Princeton University Press, 1957.

INDEX